Hormones: A Very Short Introduction

VERY SHORT INTRODUCTIONS are for anyone wanting a stimulating and accessible way in to a new subject. They are written by experts, and have been translated into more than 40 different languages.

The Series began in 1995, and now covers a wide variety of topics in every discipline. The VSI library now contains over 350 volumes—a Very Short Introduction to everything from Psychology and Philosophy of Science to American History and Relativity—and continues to grow in every subject area.

Very Short Introductions available now:

ACCOUNTING Christopher Nobes
ADVERTISING Winston Fletcher
AFRICAN HISTORY John Parker and Richard Rathbone
AFRICAN RELIGIONS Jacob K. Olupona
AGNOSTICISM Robin Le Poidevin
ALEXANDER THE GREAT Hugh Bowden
AMERICAN HISTORY Paul S. Boyer
AMERICAN IMMIGRATION David A. Gerber
AMERICAN LEGAL HISTORY G. Edward White
AMERICAN POLITICAL PARTIES AND ELECTIONS L. Sandy Maisel
AMERICAN POLITICS Richard M. Valelly
THE AMERICAN PRESIDENCY Charles O. Jones
ANAESTHESIA Aidan O'Donnell
ANARCHISM Colin Ward
ANCIENT EGYPT Ian Shaw
ANCIENT GREECE Paul Cartledge
THE ANCIENT NEAR EAST Amanda H. Podany
ANCIENT PHILOSOPHY Julia Annas
ANCIENT WARFARE Harry Sidebottom
ANGELS David Albert Jones
ANGLICANISM Mark Chapman
THE ANGLO-SAXON AGE John Blair
THE ANIMAL KINGDOM Peter Holland
ANIMAL RIGHTS David DeGrazia
THE ANTARCTIC Klaus Dodds
ANTISEMITISM Steven Beller
ANXIETY Daniel Freeman and Jason Freeman
THE APOCRYPHAL GOSPELS Paul Foster
ARCHAEOLOGY Paul Bahn
ARCHITECTURE Andrew Ballantyne
ARISTOCRACY William Doyle
ARISTOTLE Jonathan Barnes
ART HISTORY Dana Arnold
ART THEORY Cynthia Freeland
ASTROBIOLOGY David C. Catling
ATHEISM Julian Baggini
AUGUSTINE Henry Chadwick
AUSTRALIA Kenneth Morgan
AUTISM Uta Frith
THE AVANT GARDE David Cottington
THE AZTECS Davíd Carrasco
BACTERIA Sebastian G. B. Amyes
BARTHES Jonathan Culler
THE BEATS David Sterritt
BEAUTY Roger Scruton
BESTSELLERS John Sutherland
THE BIBLE John Riches
BIBLICAL ARCHAEOLOGY Eric H. Cline
BIOGRAPHY Hermione Lee
THE BLUES Elijah Wald
THE BOOK OF MORMON Terryl Givens
BORDERS Alexander C. Diener and Joshua Hagen
THE BRAIN Michael O'Shea
THE BRITISH CONSTITUTION Martin Loughlin
THE BRITISH EMPIRE Ashley Jackson
BRITISH POLITICS Anthony Wright
BUDDHA Michael Carrithers
BUDDHISM Damien Keown
BUDDHIST ETHICS Damien Keown

CANCER Nicholas James
CAPITALISM James Fulcher
CATHOLICISM Gerald O'Collins
CAUSATION Stephen Mumford and Rani Lill Anjum
THE CELL Terence Allen and Graham Cowling
THE CELTS Barry Cunliffe
CHAOS Leonard Smith
CHILDREN'S LITERATURE Kimberley Reynolds
CHINESE LITERATURE Sabina Knight
CHOICE THEORY Michael Allingham
CHRISTIAN ART Beth Williamson
CHRISTIAN ETHICS D. Stephen Long
CHRISTIANITY Linda Woodhead
CITIZENSHIP Richard Bellamy
CIVIL ENGINEERING David Muir Wood
CLASSICAL LITERATURE William Allan
CLASSICAL MYTHOLOGY Helen Morales
CLASSICS Mary Beard and John Henderson
CLAUSEWITZ Michael Howard
CLIMATE Mark Maslin
THE COLD WAR Robert McMahon
COLONIAL AMERICA Alan Taylor
COLONIAL LATIN AMERICAN LITERATURE Rolena Adorno
COMEDY Matthew Bevis
COMMUNISM Leslie Holmes
COMPLEXITY John H. Holland
THE COMPUTER Darrel Ince
THE CONQUISTADORS Matthew Restall and Felipe Fernández-Armesto
CONSCIENCE Paul Strohm
CONSCIOUSNESS Susan Blackmore
CONTEMPORARY ART Julian Stallabrass
CONTEMPORARY FICTION Robert Eaglestone
CONTINENTAL PHILOSOPHY Simon Critchley
CORAL REEFS Charles Sheppard
COSMOLOGY Peter Coles
CRITICAL THEORY Stephen Eric Bronner
THE CRUSADES Christopher Tyerman
CRYPTOGRAPHY Fred Piper and Sean Murphy
THE CULTURAL REVOLUTION Richard Curt Kraus
DADA AND SURREALISM David Hopkins
DARWIN Jonathan Howard
THE DEAD SEA SCROLLS Timothy Lim
DEMOCRACY Bernard Crick
DERRIDA Simon Glendinning
DESCARTES Tom Sorell
DESERTS Nick Middleton
DESIGN John Heskett
DEVELOPMENTAL BIOLOGY Lewis Wolpert
THE DEVIL Darren Oldridge
DIASPORA Kevin Kenny
DICTIONARIES Lynda Mugglestone
DINOSAURS David Norman
DIPLOMACY Joseph M. Siracusa
DOCUMENTARY FILM Patricia Aufderheide
DREAMING J. Allan Hobson
DRUGS Leslie Iversen
DRUIDS Barry Cunliffe
EARLY MUSIC Thomas Forrest Kelly
THE EARTH Martin Redfern
ECONOMICS Partha Dasgupta
EDUCATION Gary Thomas
EGYPTIAN MYTH Geraldine Pinch
EIGHTEENTH-CENTURY BRITAIN Paul Langford
THE ELEMENTS Philip Ball
EMOTION Dylan Evans
EMPIRE Stephen Howe
ENGELS Terrell Carver
ENGINEERING David Blockley
ENGLISH LITERATURE Jonathan Bate
ENTREPRENEURSHIP Paul Westhead and Mike Wright
ENVIRONMENTAL ECONOMICS Stephen Smith
EPIDEMIOLOGY Rodolfo Saracci
ETHICS Simon Blackburn
ETHNOMUSICOLOGY Timothy Rice
THE ETRUSCANS Christopher Smith
THE EUROPEAN UNION John Pinder and Simon Usherwood
EVOLUTION Brian and Deborah Charlesworth
EXISTENTIALISM Thomas Flynn
THE EYE Michael Land
FAMILY LAW Jonathan Herring
FASCISM Kevin Passmore
FASHION Rebecca Arnold
FEMINISM Margaret Walters
FILM Michael Wood

FILM MUSIC Kathryn Kalinak
THE FIRST WORLD WAR
Michael Howard
FOLK MUSIC Mark Slobin
FOOD John Krebs
FORENSIC PSYCHOLOGY David Canter
FORENSIC SCIENCE Jim Fraser
FOSSILS Keith Thomson
FOUCAULT Gary Gutting
FRACTALS Kenneth Falconer
FREE SPEECH Nigel Warburton
FREE WILL Thomas Pink
FRENCH LITERATURE John D. Lyons
THE FRENCH REVOLUTION
William Doyle
FREUD Anthony Storr
FUNDAMENTALISM Malise Ruthven
GALAXIES John Gribbin
GALILEO Stillman Drake
GAME THEORY Ken Binmore
GANDHI Bhikhu Parekh
GENIUS Andrew Robinson
GEOGRAPHY John Matthews and
David Herbert
GEOPOLITICS Klaus Dodds
GERMAN LITERATURE Nicholas Boyle
GERMAN PHILOSOPHY Andrew Bowie
GLOBAL CATASTROPHES Bill McGuire
GLOBAL ECONOMIC HISTORY
Robert C. Allen
GLOBAL WARMING Mark Maslin
GLOBALIZATION Manfred Steger
THE GOTHIC Nick Groom
GOVERNANCE Mark Bevir
THE GREAT DEPRESSION AND THE
NEW DEAL Eric Rauchway
HABERMAS James Gordon Finlayson
HAPPINESS Daniel M. Haybron
HEGEL Peter Singer
HEIDEGGER Michael Inwood
HERODOTUS Jennifer T. Roberts
HIEROGLYPHS Penelope Wilson
HINDUISM Kim Knott
HISTORY John H. Arnold
THE HISTORY OF ASTRONOMY
Michael Hoskin
THE HISTORY OF LIFE Michael Benton
THE HISTORY OF MATHEMATICS
Jacqueline Stedall
THE HISTORY OF MEDICINE
William Bynum
THE HISTORY OF TIME
Leofranc Holford-Strevens
HIV/AIDS Alan Whiteside
HOBBES Richard Tuck
HORMONES Martin Luck
HUMAN EVOLUTION
Bernard Wood
HUMAN RIGHTS Andrew Clapham
HUMANISM Stephen Law
HUME A. J. Ayer
HUMOUR Noël Carroll
THE ICE AGE Jamie Woodward
IDEOLOGY Michael Freeden
INDIAN PHILOSOPHY Sue Hamilton
INFORMATION Luciano Floridi
INNOVATION Mark Dodgson and
David Gann
INTELLIGENCE Ian J. Deary
INTERNATIONAL MIGRATION
Khalid Koser
INTERNATIONAL RELATIONS
Paul Wilkinson
INTERNATIONAL SECURITY
Christopher S. Browning
ISLAM Malise Ruthven
ISLAMIC HISTORY Adam Silverstein
ITALIAN LITERATURE Peter Hainsworth
and David Robey
JESUS Richard Bauckham
JOURNALISM Ian Hargreaves
JUDAISM Norman Solomon
JUNG Anthony Stevens
KABBALAH Joseph Dan
KAFKA Ritchie Robertson
KANT Roger Scruton
KEYNES Robert Skidelsky
KIERKEGAARD Patrick Gardiner
THE KORAN Michael Cook
LANDSCAPE ARCHITECTURE
Ian H. Thompson
LANDSCAPES AND
GEOMORPHOLOGY
Andrew Goudie and Heather Viles
LANGUAGES Stephen R. Anderson
LATE ANTIQUITY Gillian Clark
LAW Raymond Wacks
THE LAWS OF THERMODYNAMICS
Peter Atkins
LEADERSHIP Keith Grint
LINCOLN Allen C. Guelzo
LINGUISTICS Peter Matthews

LITERARY THEORY Jonathan Culler
LOCKE John Dunn
LOGIC Graham Priest
MACHIAVELLI Quentin Skinner
MADNESS Andrew Scull
MAGIC Owen Davies
MAGNA CARTA Nicholas Vincent
MAGNETISM Stephen Blundell
MALTHUS Donald Winch
MANAGEMENT John Hendry
MAO Delia Davin
MARINE BIOLOGY Philip V. Mladenov
THE MARQUIS DE SADE John Phillips
MARTIN LUTHER Scott H. Hendrix
MARTYRDOM Jolyon Mitchell
MARX Peter Singer
MATHEMATICS Timothy Gowers
THE MEANING OF LIFE Terry Eagleton
MEDICAL ETHICS Tony Hope
MEDICAL LAW Charles Foster
MEDIEVAL BRITAIN John Gillingham and Ralph A. Griffiths
MEMORY Jonathan K. Foster
METAPHYSICS Stephen Mumford
MICHAEL FARADAY Frank A. J. L. James
MICROECONOMICS Avinash Dixit
MODERN ART David Cottington
MODERN CHINA Rana Mitter
MODERN FRANCE Vanessa R. Schwartz
MODERN IRELAND Senia Pašeta
MODERN JAPAN Christopher Goto-Jones
MODERN LATIN AMERICAN LITERATURE Roberto González Echevarría
MODERN WAR Richard English
MODERNISM Christopher Butler
MOLECULES Philip Ball
THE MONGOLS Morris Rossabi
MORMONISM Richard Lyman Bushman
MUHAMMAD Jonathan A. C. Brown
MULTICULTURALISM Ali Rattansi
MUSIC Nicholas Cook
MYTH Robert A. Segal
THE NAPOLEONIC WARS Mike Rapport
NATIONALISM Steven Grosby
NELSON MANDELA Elleke Boehmer
NEOLIBERALISM Manfred Steger and Ravi Roy
NETWORKS Guido Caldarelli and Michele Catanzaro
THE NEW TESTAMENT Luke Timothy Johnson
THE NEW TESTAMENT AS LITERATURE Kyle Keefer
NEWTON Robert Iliffe
NIETZSCHE Michael Tanner
NINETEENTH-CENTURY BRITAIN Christopher Harvie and H. C. G. Matthew
THE NORMAN CONQUEST George Garnett
NORTH AMERICAN INDIANS Theda Perdue and Michael D. Green
NORTHERN IRELAND Marc Mulholland
NOTHING Frank Close
NUCLEAR POWER Maxwell Irvine
NUCLEAR WEAPONS Joseph M. Siracusa
NUMBERS Peter M. Higgins
NUTRITION David A. Bender
OBJECTIVITY Stephen Gaukroger
THE OLD TESTAMENT Michael D. Coogan
THE ORCHESTRA D. Kern Holoman
ORGANIZATIONS Mary Jo Hatch
PAGANISM Owen Davies
THE PALESTINIAN-ISRAELI CONFLICT Martin Bunton
PARTICLE PHYSICS Frank Close
PAUL E. P. Sanders
PENTECOSTALISM William K. Kay
THE PERIODIC TABLE Eric R. Scerri
PHILOSOPHY Edward Craig
PHILOSOPHY OF LAW Raymond Wacks
PHILOSOPHY OF SCIENCE Samir Okasha
PHOTOGRAPHY Steve Edwards
PHYSICAL CHEMISTRY Peter Atkins
PLAGUE Paul Slack
PLANETS David A. Rothery
PLANTS Timothy Walker
PLATO Julia Annas
POLITICAL PHILOSOPHY David Miller
POLITICS Kenneth Minogue
POSTCOLONIALISM Robert Young
POSTMODERNISM Christopher Butler
POSTSTRUCTURALISM Catherine Belsey

PREHISTORY Chris Gosden
PRESOCRATIC PHILOSOPHY Catherine Osborne
PRIVACY Raymond Wacks
PROBABILITY John Haigh
PROGRESSIVISM Walter Nugent
PROTESTANTISM Mark A. Noll
PSYCHIATRY Tom Burns
PSYCHOLOGY Gillian Butler and Freda McManus
PURITANISM Francis J. Bremer
THE QUAKERS Pink Dandelion
QUANTUM THEORY John Polkinghorne
RACISM Ali Rattansi
RADIOACTIVITY Claudio Tuniz
RASTAFARI Ennis B. Edmonds
THE REAGAN REVOLUTION Gil Troy
REALITY Jan Westerhoff
THE REFORMATION Peter Marshall
RELATIVITY Russell Stannard
RELIGION IN AMERICA Timothy Beal
THE RENAISSANCE Jerry Brotton
RENAISSANCE ART Geraldine A. Johnson
REVOLUTIONS Jack A. Goldstone
RHETORIC Richard Toye
RISK Baruch Fischhoff and John Kadvany
RIVERS Nick Middleton
ROBOTICS Alan Winfield
ROMAN BRITAIN Peter Salway
THE ROMAN EMPIRE Christopher Kelly
THE ROMAN REPUBLIC David M. Gwynn
ROMANTICISM Michael Ferber
ROUSSEAU Robert Wokler
RUSSELL A. C. Grayling
RUSSIAN HISTORY Geoffrey Hosking
RUSSIAN LITERATURE Catriona Kelly
THE RUSSIAN REVOLUTION S. A. Smith
SCHIZOPHRENIA Chris Frith and Eve Johnstone
SCHOPENHAUER Christopher Janaway
SCIENCE AND RELIGION Thomas Dixon
SCIENCE FICTION David Seed
THE SCIENTIFIC REVOLUTION Lawrence M. Principe
SCOTLAND Rab Houston
SEXUALITY Véronique Mottier
SHAKESPEARE Germaine Greer
SIKHISM Eleanor Nesbitt
THE SILK ROAD James A. Millward
SLEEP Steven W. Lockley and Russell G. Foster
SOCIAL AND CULTURAL ANTHROPOLOGY John Monaghan and Peter Just
SOCIALISM Michael Newman
SOCIOLINGUISTICS John Edwards
SOCIOLOGY Steve Bruce
SOCRATES C. C. W. Taylor
THE SOVIET UNION Stephen Lovell
THE SPANISH CIVIL WAR Helen Graham
SPANISH LITERATURE Jo Labanyi
SPINOZA Roger Scruton
SPIRITUALITY Philip Sheldrake
STARS Andrew King
STATISTICS David J. Hand
STEM CELLS Jonathan Slack
STUART BRITAIN John Morrill
SUPERCONDUCTIVITY Stephen Blundell
SYMMETRY Ian Stewart
TEETH Peter S. Ungar
TERRORISM Charles Townshend
THEOLOGY David F. Ford
THOMAS AQUINAS Fergus Kerr
THOUGHT Tim Bayne
TIBETAN BUDDHISM Matthew T. Kapstein
TOCQUEVILLE Harvey C. Mansfield
TRAGEDY Adrian Poole
THE TROJAN WAR Eric H. Cline
TRUST Katherine Hawley
THE TUDORS John Guy
TWENTIETH-CENTURY BRITAIN Kenneth O. Morgan
THE UNITED NATIONS Jussi M. Hanhimäki
THE U.S. CONGRESS Donald A. Ritchie
THE U.S. SUPREME COURT Linda Greenhouse
UTOPIANISM Lyman Tower Sargent
THE VIKINGS Julian Richards
VIRUSES Dorothy H. Crawford

WITCHCRAFT Malcolm Gaskill
WITTGENSTEIN A. C. Grayling
WORK Stephen Fineman
WORLD MUSIC Philip Bohlman
THE WORLD TRADE ORGANIZATION Amrita Narlikar
WRITING AND SCRIPT Andrew Robinson

Available soon:

GENES Jonathan Slack
GOD John Bowker
KNOWLEDGE Jennifer Nagel
CONFUCIANISM Daniel K. Gardner
THEATRE Marvin Carlson

For more information visit our website

www.oup.com/vsi/

Martin Luck

HORMONES

A Very Short Introduction

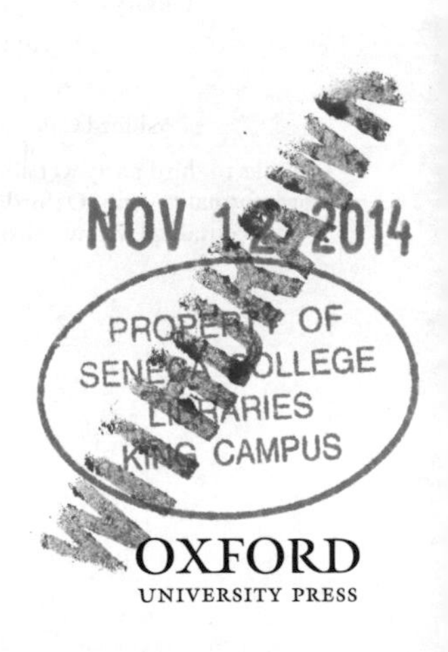

OXFORD
UNIVERSITY PRESS

OXFORD
UNIVERSITY PRESS

Great Clarendon Street, Oxford, OX2 6DP,
United Kingdom

Oxford University Press is a department of the University of Oxford.
It furthers the University's objective of excellence in research, scholarship,
and education by publishing worldwide. Oxford is a registered trade mark of
Oxford University Press in the UK and in certain other countries

First edition published in 2014
Impression: 1

Published in the United States of America by Oxford University Press
198 Madison Avenue, New York, NY 10016, United States of America

British Library Cataloguing in Publication Data
Data available

Library of Congress Control Number: 2014932709

ISBN 978-0-19-967287-5

Printed in Great Britain by
Ashford Colour Press Ltd, Gosport, Hampshire

For Jacob
whose life will be completely different from mine

Contents

Acknowledgements

This little book would not have appeared without the help of Janine Luck, Bas Haynes, Sue and Roger Golds, Louise Dunford, Latha Menon, Emma Ma, an unidentified manuscript consultant, and the three men of Edradour.

Thank you.

Chapter 1

Hormones, history, and the shoulders of giants

A matter of size

The parish register of St Anne's Church, Sutton Bonington, a Leicestershire village a couple of miles from where this book is being written, records the following death: '1773 Feb. 7th William Rice [or Riste], aged 30, seven feet four inches high.' According to village history, this 'Giant of All England' was exhibited about the country and George III gave him a suit of scarlet silk. He reputedly grew by six inches each year between the ages of 14 and 21. He was buried in the chancel of the church, his corpse being carried by eight men with eight maids bearing the pall. Upwards of 500 people attended. No sign of his grave remains but there was once a peg high up on the chancel wall on which he hung his hat.

One's view of William depends on one's interests. Village folk are presumably grateful to have a noteworthy character in their heritage, whilst historians might check the veracity of the facts reported about him. To sociologists he represents minor 18th century celebrity, whilst psychologists may wonder how he coped with constant attention and royal interest. Engineers might calculate the loads on the arms of the coffin carriers. Compilers of lists would find that although William was very tall, he holds no record: Robert Pershing Wadlow of Alton, Illinois who died in

1940 aged 22 years was 2.72 metres or 8 feet 11 inches and would have hit his shoulder on William's hat peg. Statisticians might be interested to know whether taller people have shorter lives. Doctors would want to ameliorate any height-related illnesses and seek ways of preventing the condition in others.

Anyone interested in hormones would immediately want to know if William, and indeed Robert, suffered from too much or too little of something. Many hormones influence growth and although we cannot be certain why these individuals grew so tall, we can consider some possibilities.

From being an average sized teenager, William seems to have experienced an unusually rapid growth spurt, nearly doubling his length in about seven years. Growth hormone (GH), which is produced by the pituitary gland, drives the growth of children and adolescents. Perhaps William's pituitary secreted an unusually large amount of GH at crucial stages in his life. This might indicate something wrong with the gland itself or that it was being overstimulated by his brain. The part of the brain called the hypothalamus controls much of what the pituitary does, including its production of GH. It both stimulates it, by the eponymous growth hormone releasing hormone (GHRH), and inhibits it, with the less obviously named Somatostatin. The balance between these influences decides the outcome.

Alternatively, William's secretion of GH could have been normal but perhaps his liver responded by secreting too much insulin-like growth factor I (IGF-I). This hormone does the operational work of the growth system, causing bones to lengthen and muscles to build. It also raises the body's rate of metabolism to keep pace with everything. Normally, IGF-I suppresses the secretions of the pituitary gland and hypothalamus. This sets up a correction system so that the amount of GH in the blood becomes self-adjusting. So yet another possibility is that William's pituitary or brain was just insensitive to IGF-I and he did have too much GH after all.

William eventually stopped growing but evidently did so rather late. Adolescents normally reach maximum height in late teen age because the sex hormones testosterone and oestrogen, produced by their increasingly active gonads, irreversibly alter the structure of limb and other long bones. A bone grows from its two end regions, called epiphyses (sing. epiphysis). These regions have cells that respond to IGF-I by dividing and producing cartilage. The cartilage gets smeared on the ends of the shaft, making it longer, and then becomes calcified into hard bone.

Sex hormones encourage this lengthening during the mid-teenage growth spurt but they eventually cause the epiphyses to fuse with the rest of the bone (called epiphyseal closure), thus preventing further elongation. So perhaps William's testes matured late and the slow build up of testosterone allowed his bones to lengthen for a little longer than usual.

What happens if there is too much IGF-I after limb bone growth is complete? This can happen if the pituitary develops a GH-producing tumour, and it produces the characteristic signs of a disease called acromegaly. The liver responds to the increased GH by producing too much IGF-I and bones in all parts of the skeleton become thicker and wider, rather than longer. The jaw and brow may become heavy. The hands, feet, and digits may become disproportionately large. In the days when men wore hats most of the time, an unexpected need for a larger one might be the first sign of this rare condition.

We have no information about William's facial features, nor about the size of the hat he threw on the chancel peg each Sunday, so we have no evidence of a tumour. Nor can we guess at his testicular development (the female attention at his funeral notwithstanding). There are other hormones that influence body growth, including cortisol (produced by the adrenal glands) and thyroxine (from the thyroid gland). Unusual amounts of these may have contributed to William's condition, although in that case history might have recorded some other characteristic features to give us a clue.

We cannot be sure why they grew so fast but William Rice, Robert Wadlow, and other giants represent one end of the bell curve of human heights. At the other end, there are exceptionally short individuals whose limited stature can also have a hormonal explanation. Height is partly an inherited characteristic but its genetics are extremely complex. No single gene has an overriding influence. In any case, genes do not determine characteristics—they just provide opportunities for other factors to act. Our friend William, however much GH or IGF-I he had in his blood, would not have grown so tall had his diet not given him sufficient energy and protein to do so.

We notice very tall and very short individuals because they are exceptions. Yet each of us has a unique body conformation, encompassing much more than just our height. This reflects our individual biological history. Nutrition and disease, especially during childhood and adolescence, have a large part to play. Height, weight, and aspects of health are also affected by the health, nutrition, and age of our mother during pregnancy, and probably by the circumstances of our grandmothers during their pregnancies too. Thus we inherit our bodies in an environmental as well as a genetic sense. Unusual amounts of hormones, occurring for reasons that may be ultimately unclear, impose further variations and contribute to our individuality.

Even though individual extremes of height seldom have a simple genetic cause, there are some examples of clearly inherited height difference. The Efe of Northeastern Democratic Republic of Congo are the shortest of any human population, with an average male height of 143 cm (4ft 8in). The height differential when compared to reference humans is detectable at birth and increases up to the age of 5 years.

So we might expect the Efe to be deficient in growth-promoting hormones, but in fact the levels of GH and other hormones in their blood are not particularly unusual. Instead, it seems that their body cells are insensitive to the IGF-I on which GH action depends. This stems from a mutation in the gene that codes for the IGF-I receptor, the recognition molecule that cells must have if they are to be responsive. Even a tiny change to a gene—just one letter in the DNA code—can render it useless. Mutations to the IGF-I receptor gene are extremely rare in other populations, so geneticists speculate that short stature must be somehow advantageous to the Efe hunter-gather lifestyle. Selection has allowed the mutation to be conserved in what has traditionally been a relatively isolated human group.

IGF-I regulates many aspects of metabolism in normal life and is also essential to reproduction. One therefore wonders how the Efe thrive. As its name suggests, IGF-I belongs to a family of hormones that includes insulin. Insulin has its own receptors on many cells of the body and these would be unaffected by the gene mutation. At a push, IGF-I can work through insulin's receptors although much less efficiently than it does through its own. So perhaps Efe IGF-I does work after all—not well enough to support growth but sufficient to keep the body running in other ways and to enable the population to survive.

This account of the extremes of human growth illustrates how an understanding of hormones permeates our view of how the body works. Hormones are chemicals that coordinate bodily functions. They are secreted in one location and act in another. They move around in the blood and other fluids and their effects can last for a few seconds or several weeks. As with exceptional growth, we may investigate any unusual or unwanted condition to see if a hormone has been mistimed, misdirected, or produced in the wrong amount.

Our understanding of what hormones are, where they come from and how they work is continually expanding. It is fed by investigative procedures and experimental methods that have been extraordinarily productive and reliable. For much of recorded human history, and presumably before that, accounts of disease rested on abstract or untestable ideas associated with the planets, with gods and spirits, with ill-defined miasmas or with evocative but ultimately intangible body humours. Nowadays we reach immediately for testable and internally consistent explanations, often using them as a basis for medical treatments.

There are many things we do not yet understand about hormones, but that just encourages further research because we are reasonably sure that satisfactory explanations are waiting to be found. The concept of a hormone did not exist in William Rice's time and he must have been viewed as a disturbing and inexplicable curiosity. Hormone science is little more than a century old and its rate of advance has been rapid.

Pioneers and mavericks

The word 'hormone' was first used by the English physiologist Ernest Starling (1866–1927) on 20 June 1905 in a lecture on the chemical control of body function. It was apparently coined during a Cambridge dinner with a biologist, W. Hardy, and a Greek scholar, W. T. Vesey, using the Greek *ὁρμάω* ('ormao') meaning 'I excite' or 'I arouse'.

Starling was giving substance to the novel idea that reactions occurring in the body did not have to depend on communication provided by nerves. With his brother-in-law William Bayliss (1860–1924), Starling showed that acid injected into the small intestine of a dog caused the secretion of juices from the pancreas. This happened even when all nerve connections to the pancreas had been removed. It went against the views of leading scientists of the time, including the Russian physiologist Ivan Pavlov, that all responses of the digestive system to food and other stimuli involved the brain and nervous system.

Bayliss and Starling also found (in 1902) that pancreatic secretion could be induced by injecting into the blood an extract of the wall of the small intestine. The size of the response seemed to be related to the amount of extract injected, suggesting a quantitative as well as qualitative basis for the effect. It left no doubt that chemicals rather than nerves were responsible.

They called the active principle in their extract 'secretin' (from the verb 'to secrete', not the noun/adjective 'secret') although they had little idea of what it might be. Modern textbooks describe secretin as a small protein, made of 27 amino acids. It belongs to a family of similar chemicals, all of which are secreted into the blood by parts of the digestive system. They all have something to do with the movement of food materials along the gut and with the absorption and balancing of nutrients. Another well known member of the family is the sugar-regulating hormone glucagon. About half of the amino acids in glucagon are the same as those in secretin, although the two hormones have rather different actions in the body.

Like all pioneers, Bayliss and Starling contextualized their discoveries by building on the work of their predecessors. Some fifteen years earlier, for example, Charles-Édouard Brown-Séquard (1817–94), a Mauritian-born doctor and experimental physiologist, had injected himself with extracts of the testicles of dogs and guinea-pigs. He advocated this strange practice as a route to rejuvenation and a way of extending human life. Although it evidently didn't work (he died just five years later), it was clearly based on the notion that individual organs produced substances with body-wide actions.

In 1900 again, a Philadelphian physician Solomon Solis-Cohen (1857–1948) found that an extract of bovine adrenal glands, if taken regularly, could provide some relief from asthma. 'Adrenaline' was purified from the extracts and synthesized in the lab shortly afterwards. It delivered relief when administered by inhalation, but its effects wore off very quickly and repeated doses were required. It also had dangerous side effects including adverse reactions in the lungs, heart palpitations, and raised blood pressure.

Investigation of adrenaline (US: epinephrine) eventually led to a vast and hugely productive area of pharmacological research and drug development, including the invention of salbutamol (Ventolin) and other chemicals found in the inhalers now used to treat bronchial conditions. But strangely, it was probably not the adrenaline in Solis-Cohen's extracts that was alleviating the asthmatic symptoms. The benefits probably came from the corticosteroid hormones that are also produced by the adrenal gland. These hormones are slower acting than adrenaline, which is why repeated doses of extract were needed, but they suppress inflammation—exactly the effect needed to control chronic asthma. The anti-inflammatory effects of corticosteroids have been widely exploited in modern medicine.

The discovery of secretin and adrenaline locate the beginning of modern hormone science, rather tidily, to the very start of the 20th century. Yet that science still uses language coined previously. Undoubtedly one of the principal figures in pre-modern hormone history was the French physiologist and philosopher Claude Bernard (1813–78). He was responsible for two expressions commonly employed in discourse about hormones. The first is 'internal secretion', which he used to describe the release of sugar from the liver into the blood, a discovery that contributed to later understanding of diabetes. Although sugar (glucose) is not a hormone, the phrase was widely adopted and is now applied to hormonal and non-hormonal secretions alike.

The second of Bernard's phrases is '*milieu interieur*', or 'internal environment', expressed in the statement that 'The constancy of the internal environment is the condition for a free and independent life.' The emphasis on constancy invokes an important concept that is now encapsulated in the word 'homeostasis'. This was explicitly defined in 1932 by W. B. Cannon (1871–1945) and is the idea that the inside of the body is an essentially stable place. Disturbances (for example, to the concentration of sugar in the blood, to temperature or to acidity) happen but they get corrected. Bernard probably saw stability as an esoteric manifestation of the perfection of the human body but, as his statement indicates, he also realized its necessity for proper function.

Discovery and understanding

Although hormones were not identified until the 20th century, examples of their actions were well known and often described in considerable detail by earlier physicians. It is not hard to find records of practices, observations, and discoveries that we would now interpret as having a hormonal basis (some of them will emerge in later chapters). Hindsight allows us to re-evaluate these early accounts in the language of 'internal secretions', even though Bernard and his predecessors could not have been aware of their existence.

In 1936, an American scientist, Edward Doisy, descried 'four stages' of hormone understanding (Box 1). This was a landmark event because it rationalized the subject from an experimental

Box 1. Doisy's four stages of endocrinology

Stage 1 Recognition of the gland or organ as one producing internal secretion

Stage 2 Methods of detecting internal secretion

Stage 3 Preparation of extracts leading to a purified hormone

Stage 4 Isolation of the pure hormone, determination of its structure and its synthesis

Sex Hormones (1936) Doisy EA. Porter Lectures to the University of Kansas School of Medicine; reported in *The History of Clinical Endocrinology* (1993) Medvei VC (2nd ed.) Parthenon

point of view and initiated a unified discourse amongst scientists and doctors. Doisy's account reflected the contemporary view that hormones are secreted by discrete tissues (glands) and can be studied by extracting them. Although we now know this to be an oversimplification, many discoveries have been made using this approach. A large number of hormones have been detected by perturbing stable physiological systems and finding out which tissue extracts hold them steady.

Perhaps the most famous hormonal discovery of the 20th century is that of insulin by Frederick Banting and his medical student assistant Charles Best. It had been shown in 1889 that the pancreas was essential for the avoidance of diabetes, but the mechanism was unknown. With some difficulty, Banting persuaded his professor at the University of Toronto, John McLeod, to give him a small lab and ten dogs to test his idea that a secretion other than the digestive juices was responsible. In 1921, they made a dog diabetic by removing its pancreas and found they could halt its symptoms (thirst, weakness, rising sugar levels) by injecting an extract of pancreatic tissue. (It was the classic Bayliss and Starling experimental approach.) A bulk extract of cattle pancreases worked just as well, and they even injected themselves without serious ill effects. A biochemist, John Collip, purified the active material and McLeod gave it the name insulin. In 1922, their extract saved the life of a diabetic teenager and became the first effective treatment of the disease.

Banting and McLeod received the Nobel prize in 1923. Banting was horrified that Best was not recognized and shared his half of the prize with him. Many thought that Collip should also have been recognized and McLeod eventually shared his cash with him. The team patented their discovery but gave all their royalties to their university to fund further research.

That well known story gives a reassuring humanity to hormone science. It illustrates the roles of method, motivation, and perseverance, although other discoveries in the subject have been more serendipitous or depended on maverick flair. Just as Banting and Best were discovering insulin, a French surgeon, Serge Voronoff, was using transplanted animal testes to try to stave off the effects of ageing. His treatments were in vogue for a short time and he became very rich, even though, as with Brown-Sequard's attempts in the previous century, there was no evidence of efficacy. The idea of rejuvenation using 'Monkey glands' entered popular culture and was, somewhat perplexingly, revisited in the mainstream medical literature as late as 1991.

Voronoff believed that an active substance produced by the testes would eventually be discovered. Testosterone was isolated in 1927 by Fred Koch at the University of Chicago, although by experimenting with testicular extracts in chickens rather than humans. Injecting testosterone turns out to be a relatively inefficient way of stimulating the body, yet the anabolic steroids sometimes abused by today's athletes have almost exactly the effects that Voronoff and Brown-Sequard were looking for.

The early history of hormones and the later development of the science inevitably grew out of diseases and disorders, partly because these are the things that get recorded and partly because everyone wants a cure. In human and animal medicine, any attempt at correction, say by gland removal or hormone injection, is an experiment in understanding as well as therapy. Similar things go on in modern research labs, but with the difference that the condition being investigated is deliberately contrived and experiments are controlled and capable of being repeated.

Making measurements

For the first two-thirds of the 20th century, most hormone research used live animals. Probably about half of the hormones that we know today were discovered in that time. The majority were identified, in true Bayliss and Starling style, by mincing up glandular tissue, extracting with water or some other solvent and injecting into animals, often rodents, to see what happened. The effects might be seen on the whole animal, an organ, a tissue, or a blood chemical.

Having detected a response, the next step was to quantify it and define the active component(s) of the extract so that others would accept its existence and function. A natural spin-off at this stage would be the description of an assay, or standard measuring procedure. This could be used by other researchers to ensure comparable and reliable results but also by clinical labs needing to detect diseases and monitor treatment. Labs frequently swapped 'reference' preparations with defined activity to ensure comparability.

Vast numbers of animals were used in these procedures and required industrial-scale breeding stations. Several inbred strains of mice and rats were developed and shipped in great quantity to labs all round the world. These animals were designed to have as little individual variation as possible, so they would generate uniform results. But such work was extraordinarily expensive, in terms of skilled technical assistance as well as in the cost of the animals and their upkeep. It was also relatively insensitive and inefficient: a small amount of injected hormone quickly gets diluted in an animal's circulation and the response might take a long while to show itself. A doctor trying to treat a latter day William Rice would have to wait several weeks to find out, from the growth rate of a young rat, how much GH his patient was experiencing. For many people, the use of animals in this way also became morally unacceptable.

Hormone research took a giant stride in 1960 with the description, by Rosalyn Yalow (1921–2011; Nobel prize 1977) and Solomon Berson (1919–72), of the radioimmunoassay (RIA) technique. They noticed that animal insulin given to patients to treat diabetes caused the production of antibodies. The serum containing these antibodies, coupled with a small amount of radioactively marked hormone, could be used to measure insulin concentrations in samples of blood. The method proved to be highly sensitive, inexpensive, and ideal for routine work. It involved a few animals initially—to generate the antiserum—but after that everything could be done in test tubes. Special machines and safety procedures were needed for the radioactive steps but this inconvenience was outweighed by the ability to generate lots of data very quickly.

Various adaptations of RIA followed, notably the Enzyme Linked Immunosorbent Assay (ELISA) and related methods. A further boost came with the invention of monoclonal antibodies (by Jerne, Köhler and Milstein; Nobel prize 1984), which is the technology on which dipstick pregnancy tests are based. These ways of detecting and measuring hormones are simple, fast and cheap. They are also precise, sensitive and specific (words that have particular meanings to people who measure things). But there is one caveat: although they tell you what's there and how much of it there is, none of them actually measures a hormone's effect. For that, a 'biological' method is still needed. These days, biological assays usually test the responses of cells in a culture dish rather than those of whole animals, but they are still the only way to show that a hormone does what you think it does.

Contemporary endocrinology

Although hormone science emerged from historical observations of disease, crude extracts of glands and measurements of animal responses, it finds its place today amongst the leading biological and medical specialisms. The study of hormones—what they are, what they do, where they come from, and how they work—comprises the science of endocrinology.

The term 'endocrine' was first used around 1905 following Bernard's descriptions of internal secretions. 'Crine' means 'secretion' and the 'endo' prefix indicates that those secretions take place *inside* the body. Physiologists view the body as a long, convoluted tube (the gut or gastrointestinal tract) surrounded by interesting cells, tissues, and organs. Looked at this way the central channel of the gut through which the food passes, called the lumen, is 'outside' the body. So too is the surface of the skin. Nearly everything else (circulatory system, muscle, bone, fat, brain, kidney, gonads, lungs...) is 'inside'. This view is more logical than it might seem, for the gut lumen is the first recognizable body structure to develop after conception, turning a small ball of cells into a multi-layered embryo.

Thus to a physiologist, materials that are deposited in the gut (saliva, stomach juices, intestinal fluid, digestive enzymes, bile) or onto to the skin (sweat, tears, mucus, earwax, seminal fluid, milk) go outside the body and are called exocrine secretions. To get outside they often have to pass down tubes. These may be relatively large structures, such as tear ducts and sweat pores, or just tiny channels formed by cell clusters merging into outward flowing conduits, like the gastric glands of the stomach.

Hormones, in contrast, go straight into the fluid bathing the cells that secrete them; there are no channels to direct the flow. From this extra-cellular fluid, they cross the walls of nearby capillaries and enter the blood. Because no tubes are involved, glands that secrete hormones are sometimes described as ductless.

To give a contemporary account of hormones, we will need to start by describing in some detail how they work. This will allow us to locate them accurately within the major physiological networks of the body. We will make connections with the nervous system and also delve into bones, kidneys, hearts, and intestines. Homeostasis will feature prominently but so will biochemistry, genetics, and simple physics. We will speak of molecules but also of whole bodies and the environments in which they live.

As with all biology, our understanding will be deepened when we place hormones within the context of evolution. However, enlightenment will only emerge if we are prepared to challenge some hidden assumptions and review our terminology. We will need to accept, tacitly, that man is just another animal but also allow medicine, pharmacology, and other human ingenuities to provide instructive examples. Yet that brutal acceptance must not be stretched too far lest we fail to acknowledge matters of morality and social conscience.

We can do some astonishing things with hormones, but only if society gives consent.

Chapter 2
How hormones work

What is a hormone?

A hormone is a chemical signal that enables an event in one part of the body to have an effect somewhere else. Bodies are big, complicated biological units but they are basically collections of cells. Cells need coordination to work properly. They need to grow, move energy around, make useful materials, keep warm, adjust their water content, expel waste, protect themselves, reproduce, and die.

All these things need to happen just at the right moment. It would be hard to imagine a body in which these processes happened chaotically or without coordination. Even apparently very simple organisms, including single cells like amoeba, depend on events happening in the right order and at the right time. Larger animals comprised of billions of cells owe their survival to keeping lots of massively complex processes coordinated for the duration of a lifespan. The whole is undoubtedly more than the sum of the parts.

Hormones make up one of the two great physiological control systems—the other is the nervous system—that keep the functions of the body working together. Places in the body that produce hormones are called endocrine glands or organs and are often thought of as distinct physiological locations. For example, it is easy to point to the testis, thyroid gland, or pancreas and list the hormones they produce. This is a convenient way of looking at things, and later chapters of this book will consider many of the organs involved, but it is an oversimplification. Such glands just happen to be particularly dense collections of hormone-secreting cells. They seldom produce just one hormone and they always contain cells with lots of other functions.

Many hormones come not from discrete glands but from cells distributed around the body. The gut (the digestive tract or alimentary canal) is familiar for its digestive, secretory, and absorptive functions but has an abundance of hormone-secreting cells distributed along its length. It is probably the busiest endocrine organ of all, even if we do not normally think of it that way. Some of its hormones control the inexorable passage of food material and regulate digestive and absorptive processes. Others work in the brain to control appetite and yet others influence blood sugar levels. A fascinating aspect of many gut hormones is that they fall neatly into family groups and are similar to hormones produced by the brain and other parts of the body.

Another large but distributed hormone producer is the endothelium. This is the layer of cells that lines every part of the circulatory system including veins, arteries, capillaries and the heart. Endothelial cells support immunity, help to clot blood when it is leaking, stop blood clotting when it should not, mend wounds, regulate water movement, transport chemicals, and play many other roles in health and disease. But they also secrete hormones that regulate blood pressure, change the blood flow to various tissues, cause inflammation, control the manufacture of red blood cells, and influence tissue and organ development.

Action at a distance

Hormones can act close to where they are made or a long way away. Distance is not a problem: there just needs to be a fluid, such as blood, connecting the site of secretion with the site of action. When they enter the blood, they are swept off to all parts of the body, reaching everywhere the blood goes. But they may also act locally, within the vicinity of the extra-cellular fluid.

For example, insulin is secreted into the blood by cells in the pancreas when blood sugar levels rise. It reaches the liver and muscles and makes them take up some sugar for storage or energy provision. At the same time, insulin affects cells in the pancreas right next to the ones that secreted it, stimulating them to produce other hormones. These hormones stop blood sugar from falling too far and also inhibit further insulin secretion. As a result, the insulin response is exactly adjusted to the amount by which the sugar concentration has risen. This combination of local and distant effects is one reason why sugar levels stay remarkably constant in most people most of the time.

Why does the cry of a new baby cause its mother to express milk? Because signals conveyed from her ears cause her brain to secrete the hormone oxytocin into the blood. The blood travels to cells in the breast and the oxytocin makes them contract, squeezing out the milk. But the blood takes the hormone to other places as well. She probably won't mention it, but the mother might notice stirrings in her womb whilst she feeds her baby: oxytocin makes muscles contract there as well. In another part of her brain, oxytocin has emotional effects and encourages her to bond with her baby.

As these examples illustrate, hormones provide cells with action at a distance but their distribution by blood is indiscriminate. Whether a hormone has an effect in a particular place depends on whether the distant cell has a receptor for it. A hormone may act on a limited cluster of cells, or just on one type of cell within a mixed tissue—it depends entirely on where the receptors are located. Sometimes cells all around the body may be affected, resulting in a large amplification of the initial signal. Our new mother might be stimulated to release milk if her baby suckles at one breast. Because oxytocin goes to her other breast as well, she might be surprised to find herself lactating in stereo.

Hormones are signals; it is incorrect to describe them as messages. Emails, letters, and Morse transmissions are messages because they contain encoded information that has to be read and interpreted by the receiver. Hormone molecules certainly come in an endless variety of chemical forms, but each form has a fixed shape and structure that defines its information content. Thus hormones work by being there and being recognized, not by carrying hidden data. A liver cell recognizes the presence of insulin and responds by taking glucose from the blood to store as glycogen: it does not need to decipher a message hidden within the hormone.

What *does* vary is the amount of hormone produced, how often it is secreted, and how long it lingers in the blood. These variables adjust the strength, frequency, and duration of the communication and *do* alter the informational value of the signal being transmitted. They make a big difference to the way responding cells behave. Cells also change their receptivity to hormones, for example during embryonic development, at different times of day, at different periods of the year, as the body gets older or over the reproductive cycle. They can also be turned down or off in the face of overexposure. Many hormones, including those that regulate growth and reproduction, are secreted in bursts rather than continuously—intervals of 90 minutes are typical—giving the cells they influence a chance to recover.

Cells that respond to particular hormones are called target cells. This name is misleading because it suggests that secreting cells produce their hormones with a purpose and that the hormones know where they are going. Despite this teleological error, a target cell for a hormone is defined as one with receptors to detect it. The interaction of a hormone with its receptor initiates a chain of events inside the cell that eventually alters its behaviour. The cell may change the proteins it makes, adjust its metabolism, adopt a different shape, divide, or even die. In many cases, it may respond by secreting hormones itself, raising the possibility of complex and intriguing chains of events whose outcome can be hard to predict.

A target tissue is one that contains a large number of target cells. Sometimes such tissues are easy to spot: adrenaline sets the heart racing and may loosen the gut, while prolactin will stimulate the synthesis of milk and can suppress the ovary. In other cases individual targets are less easy to define: thyroid hormones will raise the metabolic rate of practically every cell in the body, while growth hormone (GH) needs the intervention of another hormone (IGF-1) and its receptors to stimulate growth.

Connections with the nervous system and brain

Not surprisingly, there are strong, multi-functional connections between the nervous and endocrine systems. As well as only operating inside the body, hormones are only produced in response to internal stimuli. Events happening outside the body (for example, social interactions, changes in day length, the noise of an explosion, the smell of frying bacon) may result in hormones being secreted but only as indirect effects. There must first be sensory perception through sight, hearing, smell, taste, or touch, followed by the mediation of the central nervous system. The myriad interconnections between the nervous and endocrine systems mean that hormone production is continually modified by external events.

When nerves are stimulated they release chemicals called neurotransmitters. Examples are acetylcholine, serotonin, noradrenaline (US: norepinephrine), and dopamine (Box 2).

These affect adjacent nerves, muscles or other tissues but they can also be released into the blood and become hormones. Many of the hormones made by other tissues are identical or very similar to neurotransmitters, and some physiologists believe that the entire endocrine system evolved from the nerve-type secretions.

On top of all this, the brain releases its own hormones and has complex cycles of such activity. The parts of the endocrine system connected to the brain are best understood as extensions of the nervous system, releasing large quantities of neurotransmitters or other nerve-derived hormones into the general circulation. There is a distinct branch of hormone science, called neuroendocrinology, which seeks to understand how all this fits together.

Good examples of the neuro–endocrine connection are the posterior lobe of the pituitary gland, which is a downward protrusion of the base of the brain, and the adrenal medullas, which are the central parts of the two adrenal glands lying on top of the kidneys. Amongst other things, these organs take part in reflexes in which nerve activity leads automatically to sudden and unstoppable hormone-driven events.

For example, the oxytocin that initiates milk release by the mammary gland is secreted from the posterior pituitary but is actually made slightly higher up in the base of the brain, in the region called the hypothalamus. Its secretion is controlled by other parts of the brain, especially those associated with reproduction and emotion, over which the mother has little or no conscious control.

Similarly adrenaline (Box 2) is secreted by the adrenal medulla under the influence of sympathetic nerves coming from the spinal cord. It causes the familiar flight/fright/fight responses associated with shock but is also responsible for some of the undesirable effects of chronic stress. Adrenaline and its related chemicals are also made in the brain itself, but this potential confusion just emphasizes the interrelatedness of nerves and hormones.

Amongst the most complex connections between the nervous and endocrine systems are those with the front part of the pituitary gland, the anterior lobe. This organ is not part of the brain or nervous system but it does have a special blood link to the hypothalamus. Tiny proteins and other chemicals are secreted by the hypothalamus into an offshoot of the blood supply to the brain, which takes them directly to cells in the anterior pituitary. Thus although these hormones are made in minute quantities they are especially effective because they avoid being diluted in the body's general circulation. This fine control of anterior pituitary cells by the brain is one of the main ways in which the central nervous system has overall control of how some major hormone systems operate.

Hormone systems

Links between the brain and the endocrine system make up the top level of some complex hormonal systems called axes. These are systems that regulate essential processes associated with reproduction, growth, development, and metabolism.

Box 2. Tyrosine derivatives

Hormones made from tyrosine

Two groups of hormones are made from the amino acid tyrosine, which is a common constituent of proteins. The catecholamines are secreted by the hypothalamus and the adrenal medulla, and also made in the brain and sympathetic nervous system. They illustrate the close link between the endocrine and nervous systems. The hormones made by the thyroid gland have completely different actions and are unique in having iodine atoms in their structure. Each is made from two tyrosines.

Box 2. Continued

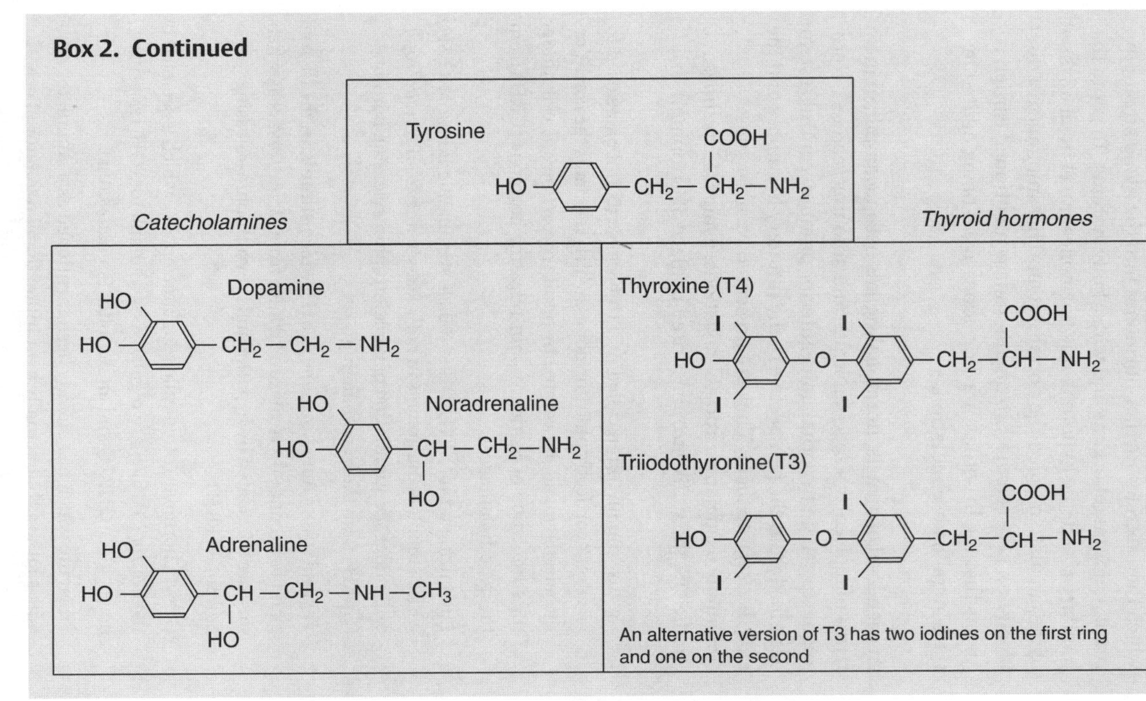

The hypothalamus sends hormonal signals to the pituitary gland, which then secretes further hormones into the blood stream to reach major target organs and wider body systems. To get to the pituitary gland, the hypothalamic hormones travel in an offshoot of the circulation called the portal system. This amounts to a direct blood link between the two organs and, as mentioned earlier, it means that the hypothalamic hormones avoid being diluted in the general body circulation.

At each level of an axis, hormonal signals become amplified in terms of molecular size, complexity, and amounts secreted. Here are the five main hypothalamic–anterior pituitary control systems, described from the perspective of the pituitary hormones and the *cells* that secrete them. They are presented with the most commonly used names and abbreviations, using UK spellings. (In the US, the suffix '-tropin' is used in place of '-trophin'.)

1. Corticotrophin (often called ACTH, because its full name used to be adrenocorticotrophic hormone), which regulates the cortex of the adrenal gland, is secreted by pituitary *corticotroph* cells under the stimulation of corticotrophin-releasing hormone (CRH) from the hypothalamus.
2. Luteinizing hormone (LH) and follicle stimulating hormone (FSH), which stimulate the gonads in both sexes, come from *gonadotroph* cells under the stimulation of gonadotrophin-releasing hormone (GnRH) from the hypothalamus.
3. Thyroid stimulating hormone (TSH), which stimulates the thyroid gland to produce thyroxine and comes from *thyrotroph* cells when they are stimulated by hypothalamic thyrotrophin-releasing hormone (TRH).
4. Growth hormone (often called somatotrophin), which is secreted by *somatotrophs* which are regulated by a stimulator (GH-releasing hormone or GHRH) and an inhibitor (somatostatin).
5. Prolactin, which has already been mentioned as a stimulator of milk synthesis but actually a multifunctional regulator of cell

biochemistry with widespread reproductive and other effects, comes from *lactotrophs*. These pituitary cells are unusual because they will secrete their hormone without any encouragement at all and need to be suppressed most of the time. The suppressor is dopamine (sometimes called prolactin inhibiting factor, PIF; Box 2), which is a brain chemical found in many parts of the central nervous system.

These major axes are interconnected and have many other regulatory features, including multi-level feedback mechanisms. They also interact with hormones not directly influenced by the brain or nervous system, such as those adjusting the concentrations of sugar and calcium in the blood. This abundant complexity accounts for the astonishingly fine and sensitive way in which hormones regulate all body systems.

All of the hormones mentioned above have a range of alternative names. Hormone nomenclature has a regrettably chequered history, partly because of the haphazard nature of discovery, including several instances of unappreciated rediscovery, and partly because a name intended to reflect secretory origin or function can turn out to be misleading when the hormone is produced at multiple sites or has different effects in different places. Well-meaning attempts to rationalize the nomenclature have been largely unsuccessful, usually resulting in more names rather than fewer and more confusion rather than less.

Hormone transport: getting around

A fundamental feature of hormones that affects how they work is their ability to dissolve. Although water and fat do not generally mix, solubility can be understood by imagining a scale with water at one end and fat (or lipid) at the other. Theoretically, all chemicals dissolve to some extent everywhere along the scale. In reality each has a position where it dissolves best. Those dissolving towards the water end are called hydrophilic or 'water loving' and those that prefer lipid are lipophilic ('fat loving'). Equally good alternative words are lipophobic ('fat hating') and hydrophobic ('water hating'), respectively. These terms are comparative descriptors: the solubility of a hormone is seldom expressed in absolute measure.

Generally speaking, the more electrical charge a molecule possesses the more hydrophilic it is. Charged molecules are often referred to as polar because they have a positive end and a negative end. The polar hormones include small peptides made of a few amino acids (TRH, oxytocin), proteins (insulin, growth hormone), protein–carbohydrate complexes (LH, FSH, TSH), a vast range of hormones derived from the fatty acid arachidonic acid (prostaglandins) and small, neurotransmitter-type hormones derived from the amino acid tyrosine (adrenaline, dopamine).

The non-polar, lipophilic hormones include all the steroid hormones (oestrogen (US: estrogen), testosterone, cortisol, etc.) and other products of cholesterol such as vitamin D, together with the thyroid hormones (which are also produced from tyrosine but are quite different from adrenaline and its friends; Box 2).

The water solubility of a hormone determines how easily it enters the blood. Plasma, the straw-coloured, non-cellular part of blood, is made mostly of water and so only relatively hydrophilic molecules will dissolve in it easily to a high concentration. Hydrophobic hormones will not dissolve freely but can be carried in the plasma if they hitch a ride on something else that is soluble. Most hydrophobic hormones have a specific binding protein that does this job. For example, oestrogen and testosterone are carried on a large globular protein called sex-hormone-binding globulin, or SHBG, which is highly polar and dissolves well. In addition, all hormones, both polar and non-polar, will bind to albumin, which is the most abundant and most soluble protein in blood. Generally speaking, binding globulins carry their designated hormones in small amounts but hold on to them very tightly. Albumin is not especially choosy about what it carries, and binds hormones in large amounts but extremely loosely.

Hormone transport: survival

Binding a hormone to a protein has a further effect: it slows its destruction by enzymes and reduces its rate of loss by excretion. This extends its active life and its potential to influence its target cells. The best measure of a hormone's survival time in the blood is its half-life: the time it takes from being secreted into the blood to only half of it remaining (Box 3).

Box 3. Half life

The 'half life' of a hormone (or any other substance in the blood) is the time taken for half of it to disappear after secretion. It describes how quickly it is metabolized and excreted. The time of *half* maximal concentration is estimated because the time at which the hormone completely disappears is difficult to define precisely.

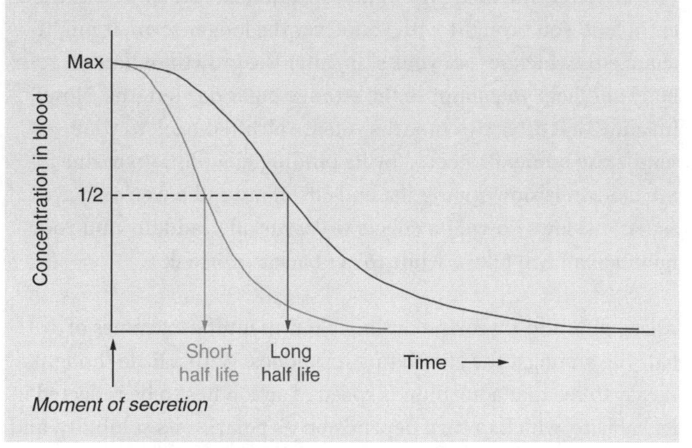

To see the importance of half-life, imagine falling feet first through the ice on a frozen lake. The shock will cause a sudden release of adrenaline from your adrenal medulla that dissolves immediately in your blood. It gets carried round your body, making lots of sugar available to your muscles, increasing your alertness and setting your heart racing. But do not worry, an enzyme waiting in the blood gets to work straight away, turning the adrenaline into inert by-products. In the minute or so it takes for you to clamber out and recover your dignity, all the adrenaline will be gone. You'll feel tired, cold, and probably angry, but you'll be safe, thanks to the drive and energy released during the adrenaline rush.

Now compare that with what would happen on an expedition to the Antarctic. It is much colder than the lake but the shock of arrival is gradual and you're well prepared with food supplies, protective clothing, and shelter. Over the first week or so, as you acclimatize, your thyroid gland becomes more active and thyroxine is secreted into your blood. Thyroxine is barely soluble by itself but has its own special protein, thyroxine-binding globulin or TBG, to help it along. The hormone slowly increases your metabolism, using up some subcutaneous fat as well as the extra food you brought with you. Over the longer term, it might change the thickness of your skin, alter the growth of hair on your body and help you adapt to the strange polar day lengths. Now imagine that after two months you are airlifted back to your temperate home. Protected by its binding globulin, thyroxine breakdown is slow, giving it a half-life of around a week. So even if secretion slows down, its effects will wear off gradually and your metabolism will take a while to get back to normal.

Adrenaline and thyroxine probably represent the extremes of half-life amongst the circulating hormones. With a little thought, it is easy to see that a hormone's speed of action has to be reflected in its half-life, which in turn depends on its polarity, its solubility, and the extent to which it is protected from breakdown and excretion.

Hormone transport: reaching the target

Hormones travel in the blood, yet blood itself doesn't bathe target cells. Blood is confined to the veins, arteries, and capillaries of the circulation (and will clot if it leaks out through a cut or bruise). The extra-cellular fluid that bathes cells is like blood but without the red cells, most of the white cells, and most of the clotting factors. The two fluids are separated by the endothelial cells of the capillaries as well as a thin layer of collagen, called the basement membrane, and sometimes a few other protein layers too.

So to reach their target cells, hormones have to move from one fluid to another and through a layer of cells and membranes, essentially reversing the journey they took when they were secreted. The membranes let water through easily and represent little if any barrier to molecules of hormone size. The endothelial cells can be a bit more discriminating as gate-keepers, especially over the rate at which they let water through, but are permissive to hormones under most circumstances. (Hormones, like other materials supplied by the blood, mostly pass *through* the endothelial cells rather than *between* them.)

So what actually causes the hormones to move? The key concept here is the basic physical process of diffusion. Diffusion means the tendency of all molecules to move around and distribute themselves in whatever fluid they happen to be dissolved in. The movement is driven by the kinetic (or heat) energy of the molecules. Given time, the distribution will even out and the concentration of molecules will be the same everywhere. But this rarely happens. Normally, there is a region of high density—for example next to the place where the hormone has been secreted—with lower concentrations elsewhere.

The interesting thing about diffusion is that the larger the difference between the two concentrations—called the concentration gradient—the faster the molecules will move from the high to the low. So the movement is rapid to start with but slows down as the concentrations even out. What really happens is that molecules move in both directions all the time but there is a *net* movement towards the low concentration. Of course, in the circulation the distribution is further enhanced by the movement and mixing of the blood as it passes around the system.

The diffusion principle applies to the blood: extra-cellular fluid junction that hormones need to negotiate, even though there are tissue layers in between. So, if a hormone is at a higher concentration in the blood than in the fluid, it will move over. This happens by diffusion alone: there is no need for a pump or any kind of active shipment from one fluid to the other.

However, the problem of solubility also crops up here. Recall that hydrophobic hormones bind tightly to binding proteins, helping them to dissolve in the blood, and that all hormones bind loosely to albumin. All such binding, whether tight or loose, is dynamic. In other words, hormone molecules are constantly jumping on and off their binding proteins. Those that have jumped off become available to pass from the blood to the tissue fluid—although their net movement is dictated as always by the concentration gradient.

For the lipophilic hormones, and for some of the more hydrophilic ones too, proteins and protein-carbohydrate complexes in the target tissues may provide an attractive harbour. They concentrate the hormones at target sites but also effectively take them out of solution. In this way the hormones can influence their target cells without being involved in the concentration gradient. This also speeds up the diffusion of more molecules out of the blood and into the tissue fluid.

Hormone action: receptors

Just as hormones vary in structure, so do their receptors. Many receptors are to be found on the surfaces of cells, anchored in the membrane, while others live in the cytoplasm and can move to the nucleus. For both types, activation provokes a cascade of further enzymic and other steps, rather like the toppling of a line of dominoes, which eventually changes the behaviour of the cell. Much of the work of pharmacologists is devoted to understanding how receptors work and designing synthetic, hormone-impersonating chemicals to activate or block them.

As with transport in the blood, solubility turns out to be a good guide to the type of receptor a hormone might have. The membrane of a cell is made of phospholipid coated on both sides with protein. One protein coat faces inwards towards the cytoplasm of the cell and the other faces outwards towards the watery extra-cellular fluid. The phospholipid which fills the sandwich is a hydrophobic environment. It forms a barrier to the entry of polar, hydrophilic hormones but no barrier at all to non-polar, lipophilic ones.

Receptors for hydrophilic hormones are proteins and they reside in the membranes of target cells. Within their structures they have some relatively hydrophobic amino acids. These hold the receptor molecule in place by automatically locating it to the phospholipid filling of the membrane sandwich. Hydrophilic amino acids on one end of the molecule poke out into the extra-cellular fluid, while hydrophilic amino acids on the other end poke into the cell cytoplasm.

Some receptors have seven hydrophobic regions or domains, causing the receptor to twist in and out of the membrane, rather in the shape of a snake. Others have a single membrane-spanning domain but operate in pairs: identical or mirror image proteins, together called a dimer. In both cases, the hormone binds to the outward facing part of the receptor molecule. This part of the receptor is called the hormone-binding domain and has an arrangement of amino acids that exactly fits the shape of the hormone. This interaction of shapes, together with complementary arrangements of electric charge, is what gives specificity or uniqueness to the interaction of a hormone and its target cell.

What happens when a hormone encounters its receptor? For a receptor of the snake-like type, the binding of the hormone produces a sudden change of shape, in much the same way as a clasp lid flips when pressed onto a jam jar. This is transmitted to the intracellular region of the receptor molecule and results in the activation of other proteins, called G-proteins, within the cell. In the case of the single membrane-spanning type of receptor, the presence of the hormone draws the two units of the dimer together, forcing them closer along the membrane. The regions of the receptor inside the cell are either enzymes themselves or able to attract enzymes in the vicinity. When brought together these enzymes, which were previously inert, activate each other.

Thus polar, hydrophilic hormones influence their target cells without entering them. In contrast, hydrophobic hormones such as steroids can pass from the extra-cellular fluid directly into the cell, taking advantage of the hydrophobic phospholipid layer of the membrane. In fact it is realistic to think of cells as being constantly invaded by all manner of hydrophobic molecules, passing in and out through the membrane without let or hindrance.

Thus receptors for hydrophobic hormones are to be found inside the cell, within the cytoplasm. As with hydrophilic hormone receptors, the molecular shapes of cytoplasmic receptors are crucial in deciding whether the cell responds to a particular hormone or not. If molecular recognition does happens, the binding of the hormone causes an identical pair of receptor molecules to stick together, forming a dimer. The whole hormone–receptor arrangement moves from the cytoplasm into the nucleus of the cell. It is now in an activated condition, and specific regions of the duplicated receptor engage with the double helical strands of DNA. This supresses or activates individual genes, thereby changing the proteins the cell makes and altering its behaviour.

Hormone action: complexity

Relatively recently it has become clear that this simple division of hormones into two groups on the basis of solubility—hydrophilic and hydrophobic—does not completely explain how many of them work. It turns out that many hydrophobic hormones have membrane receptors as well as cytoplasmic ones. Furthermore, some of them can interact directly with enzymes and other processes inside cells without binding to a separate receptor protein at all.

One result of having membrane receptors as well as intracellular ones is that some hydrophobic hormones operate on more than one time scale. The process of transcribing DNA into RNA, translating RNA into protein, and processing the final product for secretion is relatively slow (30–60 minutes at best), so a hormone working through this route takes a long time to change cell behaviour. Hydrophilic hormones, working through membrane receptors, generally have much quicker effects, measured in seconds rather than minutes. The slower mechanism was discovered first and thought for a long time to be the only hydrophobic one. Then endocrinologists started to observe impossibly fast target cell responses and had to investigate other possibilities. It is now recognized that these hormones often work through membrane receptors too, or may even avoid receptors completely.

This range of mechanisms presents a multi-hued palate of options for hormonal action. Besides teaching endocrinologists not to make assumptions, it has been fully exploited during the evolution of biological systems. Cells are exposed to a constantly changing environment of stimulators and inhibitors, reflecting the multiplicity of events happening elsewhere in the body. Variety, interaction, and overlap between mechanisms explain how they adjust to this profusion and confusion of influences.

Chapter 3
The mysteries of reproduction

There are many people alive today who owe their existence to a particularly obliging group of Italian nuns. From about 1960, retirement convents within easy reach of Rome were visited on a regular basis by tanker trucks from the pharmaceutical company Serono. The trucks were collecting urine and their unappealing mission had the blessing of none other than the pope's nephew, Don Giulio Pacelli. In fact, the Vatican was a major shareholder in Serono and Pacelli was a board member. With this infallible endorsement, the nuns presumably felt they were advancing the Lord's work and would be appropriately rewarded for their inconvenience.

Vast quantities of the precious amber fluid were processed in Serono's factories, principally by passing it through cartwheel-sized cakes of kaolin. Besides its distinctive smell, urine is a perfect culture medium for bacteria, so one hopes the factory workers were suitably recompensed too. A messy and laborious extraction procedure eventually generated a hormone preparation. It was called human menopausal gonadotrophin, and Serono christened it Pergonal.

A decade previously, Bruno Lunenberg and colleagues in Israel had discovered how the pituitary gland controls the ovary. Ovaries have fluid-filled blisters called follicles and each follicle contains an egg that can eventually be released for potential fertilization by sperm. A pituitary secretion, imaginatively called follicle-stimulating hormone (FSH), causes follicles to grow and mature. Half-way through the month, the release of an egg is triggered by another pituitary hormone called luteinizing hormone (LH).

Both FSH and LH are excreted by the kidneys more or less intact, so Lunenberg reasoned that a suitable extract of urine had the potential to treat infertility. In 1961 his team injected this into an Israeli woman who had been unable to ovulate. She fell pregnant. About 10 per cent of couples experience infertility of one kind or another, so this was a major step forward in reproductive medicine.

Serono exploited this breakthrough commercially, but what was so special about the ageing nuns? Post-menopausal women, whether religious or not, produce FSH but have unresponsive ovaries. They do not develop follicles and they therefore lack the hormones that follicles secrete. Amongst these missing hormones are a couple of proteins, called Inhibin A and Inhibin B, as well as the female sex hormone oestradiol. Before the menopause, the amounts of these hormones in the blood fluctuate over each menstrual cycle in a characteristic way: Inhibin B rises and falls during the first half (the follicular phase), while Inhibin A and oestradiol rise and fall during the second half (the luteal phase).

The effect of this cyclic hormonal dance is to hold the secretion of FSH in check. As a result, the pituitary produces just enough of it to make follicles grow—and produce their hormones—but no more. It is a classic negative feedback control system. (The origin of the name inhibin is not difficult to work out.) At the menopause, as follicle development gradually subsides and menstrual cyclicity peters out, oestrogen and inhibin secretion decline. Their levels in the blood get very low, leaving the pituitary's FSH secretion to run out of control. Over the course of four or five years, amounts of FSH in the blood can increase by as much as ten times, with correspondingly more going down the toilet. So the convent donation, undiluted by urine from cycling women or (presumably) males, was a reliably rich source of FSH. Nevertheless, it still took ten elderly nuns about ten days to produce enough for Serono to extract one therapeutic dose.

Pergonal not only stimulated the ovaries of women who couldn't produce eggs. It also proved useful in treating women with Fallopian tube problems where carefully timed development of eggs was necessary prior to IVF. Lesley Brown, mother of Louise (b. 1978) and Natalie (b. 1982), received Pergonal, as did the mothers of most other test-tube babies conceived over the next three or four decades. There is an appealing irony in the notion of elderly celibate women helping their uncloistered sisters to have babies. Even more ironic, given Pergonal's papal connections, was the Vatican's dislike of IVF and related reproductive technologies, as illustrated by its criticism of the awarding of the 2008 Nobel prize to IVF pioneer Robert Edwards.

Catastrophes

The story of Pergonal touches on some key events in the remarkable process of reproduction. The main action of the menstrual cycle happens, unseen, in between the two phases: the release of an egg suitable for fertilization. The blood levels of the two inhibins are low at this point in the cycle and the result is a remarkable change in the effect of the oestradiol. Having previously inhibited the pituitary gland, it now becomes a stimulator, increasing the secretion of the other gonadotrophin, LH. The effect of LH on a growing ovarian follicle is to make it produce yet more oestradiol. So, for a short while, more oestradiol means more LH, and more LH means more oestradiol.

Eventually, over a few hours, the ovary responds to the surging LH by producing tissue enzymes. These digest the wall of the follicle and it collapses. Its contents, including the egg and a volume of fluid, ooze out into the reproductive tract ready for the attention of any sperm that may be lurking there.

What we have just described is a positive feedback system rather than a negative one. Negative feedback mechanisms wind processes down and are good for keeping things stable, so it is not surprising that they exist all over the endocrine system. They are effective at dealing with disturbances and are part and parcel of the homeostasis, which Claude Bernard described and which physiologists take for granted.

The point about positive feedback loops is that they wind themselves up rather than down. The signals—oestradiol and LH in this example—reinforce each other but the wind-up has to end in a singular event of some kind. Thus positive loops are no good for stability but they are great for drama. An engineer would describe this as an accumulation and sudden release of energy, followed by a return to the initial state. Borrowing the terminology, the point where it becomes unstoppable and something has to give is called a 'catastrophe'.

The reproductive system makes use of several positive feedback loops. The milk ejection reflex, in which milk is squirted from the breast in response to a baby suckling or crying, is caused by oxytocin secreted from the posterior pituitary gland in response to a positive nerve signal. It is uncontrollable once it starts. Similarly, the build-up of contractions during labour depend on this kind of positive reinforcement between nerve signals and oxytocin, and end with the biggest drama of all—the birth of a baby. Ejaculation in males (a very complex process, driven mostly by a reflex in the spine) and orgasm in both sexes involves positive feedback too. Somehow it is fun to be able to describe these events as catastrophes.

Infertility and prolactin

A relatively common type of infertility in women has to do with the secretion of the anterior pituitary hormone, prolactin. Normally, the amount of prolactin secreted is quite low, but this changes after giving birth. Prolactin levels increase, stimulating the mammary gland to make milk. No one is quite sure why but these high levels also inhibit the ovary, which is why suckling and lactation can have a contraceptive effect.

Unfortunately, some women experience high levels of prolactin at other times. This prevents the monthly cycle from happening, a condition called hyperprolactinaemic amenorrhoea, and the breasts may even produce milk (a coincidence of signs described even by Hippocrates around 300 BCE). Pergonal and other gonadotrophin-based treatments were sometimes used to overcome this condition. They proved effective in stimulating the development of follicles, but it was a risky approach because the response of the ovary was unpredictable and many multiple births resulted.

The strange thing about prolactin is the way its secretion is controlled. Most pituitary hormones are only produced when they are stimulated by hormones coming from the brain. Prolactin secretion will occur without stimulation, so it is normally inhibited by the brain and its secretion depends on the removal of the inhibitor. Inhibition is more or less constant, in males as well as females, which is why most women are fertile, most men do not develop breasts, and most people are not lactating. It takes the hormonal upheavals of pregnancy and birth to switch the inhibitor off.

The inhibitor's identity was a mystery for many years and was just called 'prolactin inhibiting factor'. Researchers were mystified, partly because they expected it to be a small protein or peptide like the other brain hormones that affect the pituitary gland. But in the early 1970s it was found to be something already well known as a neurotransmitter, called dopamine. Dopamine is a similar sort of chemical to adrenaline (Box 2). It occurs in many parts of the nervous system and helps to coordinate the activity of brain cells. It is part of the system that goes wrong in Parkinson's disease, for example. The dopamine produced in the hypothalamus leaves the brain by the portal system, enters the pituitary gland and suppresses the lactotroph cells located there.

The discovery of dopamine's role led to much improved treatments for this type of infertility. A synthetic chemical called bromocriptine, which is a long-acting analogue of dopamine, was developed to suppress prolactin in women with too much. Bromocriptine in pill form became a simple treatment for women with hyperprolactinaemia who wished to conceive. The really neat thing about this was that it avoided the risk of multiple pregnancy. Lowering the level of prolactin allows the ovaries to respond to FSH and LH in the normal way rather than needing to be being artificially stimulated.

Prolactin is an unusual hormone in other ways. For one thing, although it is named for its milk stimulating role on the mammary gland, it is found in many different kinds of animal and is reckoned to have more than 300 different actions. Birds, reptiles, amphibians, and many fish have prolactin, even though none of these animals feed their young in the bizarre way that mammals do. Perhaps the nearest are pigeons, which feed their newly hatched squabs on a milk-like secretion from the skin in the crop region of the breast. That is also stimulated by prolactin and was once used as a crude way of measuring the hormone, but the process scarcely resembles mammalian lactation.

Prolactin has the general effect of controlling the movement of salts and water across membranes, especially in tissues related to the skin. That is actually what it does in the lactating breast, except that we see the result as the manufacture of milk. In migratory fish like salmon, prolactin regulates the water balance of the body as the fish move between salt and fresh water during their complex annual reproductive cycle. In birds, prolactin causes the annual feather moult—another skin-related event. In some mammals, especially rodents, prolactin directly stimulates the ovary and is needed for pregnancy. It also has a wide variety of behavioural effects, including a role in libido and feelings of sexual arousal and gratification. In development, it seems to be necessary for nerve cell formation, including the production of the myelin coat found on many axons. Several different types of cancer cells are also sensitive to prolactin.

A contemporary view of the multi-tasking prolactin is that it is as much a cytokine as a hormone. Cytokines are a diverse group of molecules involved in all kinds of communication between cells, especially those of inflammation and immunity. Some of them, like prolactin, have classical hormone-like features: they come from distinct organs and can be easily measured in the circulation. Others, such as the interleukins of the immune system, come from individual blood cells such as the macrophages and can be hard to detect until the body suffers an infection. Prolactin works on cells through membrane receptors that have the shape, structure, and action typical of those for cytokines, so perhaps this was indeed the group of signalling molecules from which it originated. It is certainly very old and its biological role has evolved in many different directions.

Getting pregnant

Here's another reproductive mystery: how does the ovary know when the uterus is pregnant? The first clue a woman gets that she might be pregnant is that her menstrual cycles stop. Because this is a change in the uterus it may seem obvious that it is connected with the presence there of a fertilized egg. But reproductive cycles, as we have seen, depend on hormones secreted by the ovary, the pituitary gland, and the brain, and it is not at all clear why conception in the uterus should influence these organs and stop the cycle in its tracks. After all, the newly implanted embryo is miniscule and embedded far away from where the hormones are being secreted.

The mechanism—labelled the maternal recognition of pregnancy—took considerable effort to work out, although it was unravelled in humans a good deal earlier than in other species. The human reproductive system is quite a bit different from those of most familiar animals, including the domestic, farm, and wild species in which humans take the closest interest. It seems that evolution has reinvented, or at least reconfigured, the mammalian reproductive wheel several times since we diverged from the ancestor we shared about 150 million years ago.

The human menstrual cycle is demarcated by the obvious event of bleeding, together with more subtle fluctuations in mood, behaviour, skin sensitivity, fluid balance, and even body temperature. For convenience, the start of menstruation is counted as day one. Ovulation happens about a fortnight later, neatly dividing the 'month' into halves.

In nearly all other mammals that cycle on a regular basis, there is no bleeding and no equal division between the phases. The convention in these animals is therefore to start counting the cycle from the day of ovulation. But this event can be tricky to detect. Dairy and pig farmers have to use subtle and mostly imprecise behavioural indicators of what's going on, so that they can arrange for mating or artificial insemination at the most propitious moment. Under natural mating conditions, the males of these species use odours, called pheromones, to detect female heat. This is much more reliable but undetectable, unfortunately, by the stockman.

The second half of the human cycle is called the luteal phase because the follicle that released its egg at ovulation turns into a dense yellow body called the corpus luteum (CL). This secretes large quantities of a steroid hormone called progesterone. Progesterone keeps the lining of the uterus in a receptive state, ready to receive a fertilized egg. It also inhibits the brain and pituitary gland from secreting the FSH and LH which support follicle development and ovulation. This is why it works as a contraceptive agent—used in the 'pill'.

The physiological difference between pregnancy and non-pregnancy is that the life of the CL, and therefore the secretion of progesterone, is extended. So the question 'How does the ovary know that the uterus is pregnant?' becomes 'How does the conceptus (the fertilized egg and its membranes) stop the CL from disappearing?'

We owe the first clues to an answer to gynaecologists and vivisectionally inclined animal scientists of the first half of the 20th century. They observed what happens to the CL of a normal cycle if the uterus is surgically removed. In humans, the answer to the question is, 'Nothing' (at least not for some time): the ovary continues to cycle between follicular and luteal phases although of course there is no menstrual evidence of this. In cows and sheep and many other animals, the opposite happens: removing the uterus causes cycles to stop. The life of the CL is extended, just as if the animal was pregnant. So it seemed that while human pregnancy interrupted the cycle and supported the CL, in many other animals it stopped the uterus from causing its decline.

The human trail led to the discovery of a unique hormone, chorionic gonadotrophin (hCG; the 'h' stands for 'human'). This is produced by the embryo within a very few days of fertilization and implantation. It is similar to LH but it quickly reaches much higher concentrations in the blood. (Cleverly distinguishing hCG from LH before the concentration goes up is what makes dipstick-type pregnancy tests reliable.) The hormone hCG swamps the ovary and rescues the CL by halting its self-destruction. The CL continues to grow, producing enough progesterone to keep the lining of the uterus in the right condition for the pregnancy to be established.

In most other animals there was no CG to be found, so working out the answer to the pregnancy question proved very tough indeed. In the early 1970s, experiments on cows and sheep showed that the cause of CL destruction at the end of each cycle, in the absence of conception, was a small, short-lived hormone called prostaglandin $F_{2\alpha}$. This was produced by the lining of the uterus but it came in tiny amounts. Moreover, it disappeared from the circulation almost immediately after secretion, as soon as the blood had passed once through the lungs. So how it could it really be affecting the ovary?

The answer was anatomical: the tiny veins carrying blood away from the uterus run very close to the small branches of artery carrying blood to the ovary; in fact they wrap around each other like layers of basketwork. These capillary vessels have very thin walls, enabling the prostaglandin to jump directly from the blood leaving the uterus into the blood heading for the ovary. In this way, it neatly avoids both inactivation in the lungs and dilution in the general circulation, reaching the ovary at an effective concentration.

The CL responds to the prostaglandin by secreting oxytocin (the same hormone as produced in the brain, but discovered much later). This goes into the general circulation, reaches the uterus and stimulates the secretion of even more prostaglandin. So yet another positive feedback loop is set up. What is its catastrophic finale? The demise of the CL and the resumption of cyclicity.

If a fertilized egg implants itself in the uterus, the secretion of prostaglandin is inhibited, and the positive feedback loop never gets going: the CL survives and its progesterone supports the pregnancy, just as it does in humans. One loose end in the mystery was how the tiny fertilized egg could do this, being many times dwarfed in size by the uterine lining. Once again, a local hormone does the job. This time it is a small protein called interferon tau, which belongs to the family of chemicals produced by the immune system during a viral attack. Pregnancy appears to be acting like a uterine infection, although it is not safe to take that analogy too far.

The human and ruminant systems for the maternal recognition of pregnancy represent easily contrasted, essentially opposite, solutions to the problem, but there are many other systems to be found. Virtually all species have their individual twists on these and other arrangements. Animals that do not cycle regularly (cats and dogs, for example) or that need the stimulus of mating to ovulate (rabbits, ferrets) provide especially interesting strategies. Like all reproductive systems, they can only be fully understood in the context of the lifestyle of the animal: the varying effects of season, the length of the reproductive lifespan, and the extent to which males and females live close or apart from each other are but a few of the essential considerations that distinguish one species from another.

'Male' and 'female' hormones

It is common to refer to the sex hormones by gender—oestrogen as the female sex hormone and testosterone as the male one. This is convenient shorthand but it is misleading: hormones are not gender-specific and the terminology simply reflects our desire to make a clear distinction between male and female individuals.

For one thing, at the chemical level, you cannot have oestrogen without testosterone: oestrogen is produced in the ovaries, the testes, and elsewhere by the action of an enzyme on testosterone (or its close partner, androstenedione; collectively such hormones are known as androgens). The manufacture of oestrogen is just the next step in the chain of hormone processing (Box 4). This step goes by the poetic name of 'aromatization' because it gives the molecule a cyclic ring structure like that in the volatile chemical benzene. The essential point is that not all the androgen gets converted in this way. Animals we call female happen to convert more of it than those we call male, but it is not the exclusive prerogative of one sex or one type of gonad.

Second, males need oestrogen and females need androgen. In both sexes, oestrogen regulates important metabolic activities including food intake, body weight and fat distribution, sugar balance, and the effectiveness of insulin. It also influences muscle function, bone structure, the cardiovascular system, water balance, and the immune system. So it would probably be fair to say that all the body's cells are in some way oestrogen-dependent or at least oestrogen-sensitive. Age-related loss of oestrogen and androgen can be a contributory factor in the development Alzheimer's disease.

Female genitalia require some androgen to develop properly, even though exposure to too much during uterine development can lead to problems later on such as polycystic ovaries and type 2 diabetes. Female libido is well known to be androgen-dependent and levels of testosterone in the blood rise just before ovulation, as the most fertile period of the menstrual cycle approaches. Testosterone may affect other aspects of female behaviour and psychology in rather subtle ways. In a recent carefully controlled study, women given small amounts of testosterone became more sceptical and less trusting of the faces of strangers, compared with women given a placebo. Despite this, the baseline testosterone concentration of the women before the experiment was no guide to their degree of trust, and nor is it an absolute guide to a woman's libido or the likelihood of sexual dysfunction.

Essentially, the picture is that the sex steroids are crucial in early development but also affect function later on. The chicken's reproductive system provides a good example of the interplay of oestrogens and androgens. As a young hen reaches the age at which it will lay eggs, its ovary secretes oestrogen. The oestrogen stimulates the liver to make the proteins for egg yolk. These circulate in the blood and are taken up by growing follicles in the ovary. Oestrogen also stimulates the growth and development of the oviduct, giving the bird the machinery to make albumen for egg white, membranes to hold the egg together, and calcium carbonate for the shell. It also makes the bird's comb and wattles grow and take on the bright coloration characteristic of the laying condition.

But all these effects of oestrogen only work if the liver, oviduct, and head skin are first primed with a little testosterone. This also comes from the ovary, made mostly from the very small follicles that have yet to grow or acquire yolk. Of course, male birds produce testosterone, so what happens if you inject oestrogen into a cockerel? Hey presto, its liver will start to make yolk and its head ornamentation will change shape. The cockerel will also alter its behaviour and, with continued treatment, the sound of its voice.

The need for androgen in the hen and the effects of oestrogen on a cockerel illustrate nicely the gender-blindness of hormones and their actions. There really is no fundamental difference between

Box 4. How steroid hormones are made

Steroid hormones are made by cells in the ovary (O), the testes (T), and three zones of the adrenal cortex (Zr, Zf, Zg*). The shaded parts of the pathway are the same in each gland. The other parts happen more, or only, in some glands rather than others. The arrows hide a lot of intermediate steps and only the main secreted hormones are illustrated. As a shorthand, endocrinologists often group steroids by the number of carbon atoms they have (C27 etc.; angles and points in the diagrams). All steroids start from cholesterol. This comes mostly from the blood, but cells can also make it. The slowest part of the whole pathway is feeding cholesterol to an enzyme in the cells' mitochondria so that they can make the first conversion product (pregnenolone, a C21). Hormones which speed up steroid production (LH, FSH, hCG, ACTH, and AngII) work by making more cholesterol available.

Box 4. Continued

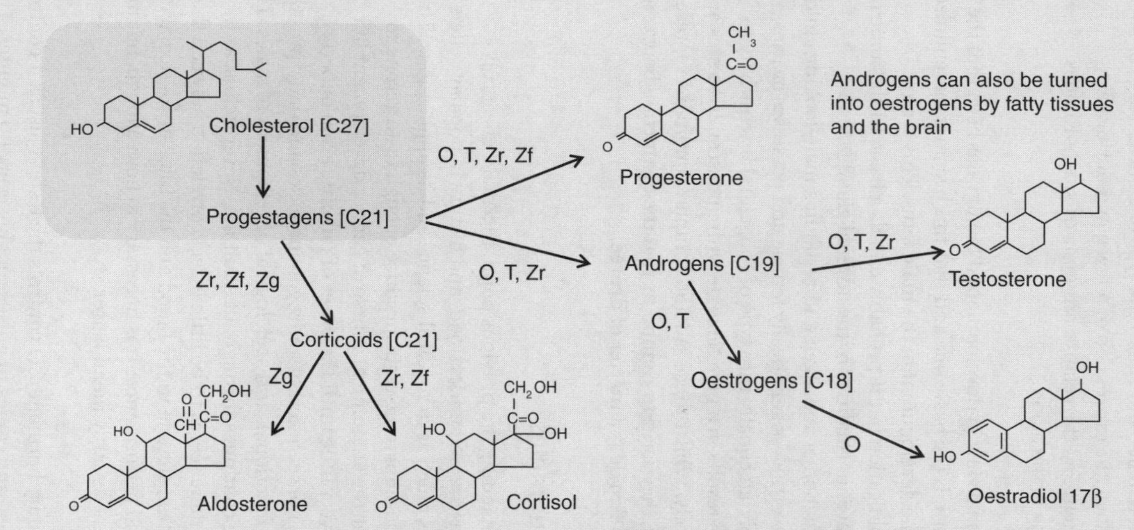

*Layers of the adrenal cortex, moving outwards:
Zr= zona reticularis, Zf= zona fasciculata, Zg= zona granulosa

This box does not show steroids made by the placenta. Vitamin D is shown in Box 6

males and females other than the type of gametes (eggs or sperm) they produce, the expression of a few genes during early development and the relative amounts of 'sex' hormones circulating in the blood.

A traditional view in developmental biology has been that the female form is the basic one and that this becomes masculinized early in the development of the male fetus. This is an oversimplification but it probably entails a basic truth. Some rare congenital abnormalities of the adrenal gland lead to overproduction of androgen and result in a male-looking child who is nevertheless genetically female and possesses ovaries. Conversely, an equally rare type of congenital insensitivity to androgen results in a genetic male with undescended testes and a female body. Whilst these sad and complex conditions represent extremes, they serve to remind us that the distinction between male and female is a matter of degree.

All about males

Notwithstanding the previous point, the 'male' effects of androgens are well understood, and feature in popular beliefs and culture as much as in medical science and its applications. The cells of the embryonic testis produce enough testosterone to determine the anatomical course of gender development but then produce very little until the onset of puberty. From puberty onwards, the secretion of androgen increases and underpins the physical, emotional, and sociological development of males. This influence continues throughout adult life. Nevertheless, as in females, there seems to be little direct correlation between the levels of testosterone in the blood and the intensity of sexual drive, libido, and performance. It is necessary to have enough, but more does not necessarily mean better.

The pituitary hormone LH controls the testes (just as it controls the ovaries) and comes in pulses roughly every 90 minutes. This pattern happens in everybody all the time, and reproduction and development depend on it. The pulses may change in size and they might speed up or slow down a little (they do this over the menstrual cycle too), but the underlying pattern is always the same. Careful experiments in animals show that each pulse of LH coincides with, or slightly follows, a pulse of gonadotrophin releasing hormone (GnRH) coming from the brain. So it is really the brain that is driving things.

There is a rare condition in boys (with the accurate but tongue-bending name hypogonadotrophic hypogonadism) in which the brain fails to get things going and puberty is delayed. This can be cured by inserting a small chemical pump under the skin that replaces the missing pulses of GnRH. The pituitary gland immediately responds with pulses of LH, and then the testes respond with normal bursts of testosterone, kick-starting near normal development. For all we know, a device like this might have limited the height of William Rice, our giant from Chapter 1.

Castration of juvenile males has long been used in animal husbandry to make stock more docile, faster growing, and more agreeable to the palate. The same technique applied to humans, in ages with different moral and social values from our own, was used variously to facilitate risk-free harem management and extraordinary singing. Carried out in adulthood, by surgery or chemical means, it has been used as a punishment, as a correction for 'deviant' behaviour, and to achieve compliance amongst soldiers and prisoners. As recently as 1952, the mathematician and cryptographer Alan Turing was subjected to chemical castration as a penal retribution for homosexual behaviour (he subsequently committed suicide).

The effect of surgical castration in humans is less clear than popularly imagined. Italian castrati were known to be able to achieve erections and orgasm and there was reportedly some social caché to be gained by any high society woman who contrived to have an affair with one. This reflects the fact that the mechanics of male reproduction are essentially testosterone independent. A small amount is needed for adequate development of the muscles and blood vessels of the penis and associated anatomy, and glands such as the adrenal cortex probably supplied this in the castrati. Erection and ejaculation have more to do with local adjustments to blood flow and spinal reflexes than with sex hormones. Prolactin, oxytocin, and a few other hormones play a supporting role but their presence in the blood after ejaculation contributes to loss of arousal and even yawning.

In contrast, sperm production and health are distinctly androgen-dependent. The sperm-producing seminiferous tubules of the testes are interspersed with Leydig cells, named after Franz Leydig (1821–1908) who first described them in 1850. These produce the testosterone and other androgens that drive the manufacture, support, and maturation of the spermatozoa, and also support accessory glands in the male reproductive system such as the prostate.

Interestingly, the development of sperm within the testis and their subsequent maturation in the epididymis prior to ejaculation depend also on progesterone, offering a further blurring of gender distinctions between hormones. Progesterone is best known as the steroid hormone produced by the ovary (the CL) during pregnancy but it is secreted in significant amounts by the testis and adrenal glands of males too. It allows the tail of the sperm to develop and generate sufficient motive force for movement through the female tract and penetration of the egg at fertilization.

Menopause and ageing

The female menopause is such a familiar, well described, and exhaustively analysed event in reproductive life history that one would imagine it to be completely understood. Yet although oestrogen secretion is known to decline as ovarian follicles cease to develop and mature, it is fair to say that much more is understood about causes than effects.

For example, the periodic hot flushes (US: hot flashes) experienced by many women have been variously attributed to the lack of oestrogen, to the elevated pituitary FSH (the reason for the convent donations), and to elevated GnRH originating from the hypothalamus (which also goes out of control for want of feedback). Yet the temperature- and sweat-regulating sites where all or any of these hormonal changes may be acting are far from understood. The emotional and skin changes that women go through at this time are similarly open to further research.

Hormone replacement therapy, whether based on synthetic oestrogen or on the equivalent hormones produced by horses (equilin and equilenin), can limit these symptoms but this merely emphasizes that the fundamental issue is loss of oestrogenic action; it does not explain at a scientific level why the symptoms occur as they do. Lack of oestrogen is certainly a direct cause of the loss of bone mineral and the consequent increased risks of bone fractures and curvature of the spine. Bone is recognized as a target for the hormone and so this symptom of menopause is well understood. But with the other symptoms, medical science usually has to fall back on the true but frustratingly vague explanation that all the body's cells are oestrogen sensitive.

The male menopause, sometimes caused andropause, is less well defined and more controversial. Either some men suffer and some do not, or it is a consequence of gradual male ageing, which varies in intensity or it is the result of convenient autosuggestion. The explanation would be simple if all men experienced a measurable decline of testosterone secretion with age, but they do not. In any case, as pointed out earlier, amounts of androgen in the blood do not have a simple relationship with the hormone's anticipated actions.

The reason for this is that the effectiveness of testosterone and similar steroids depends on what happens in the target cell as much as on the concentration of the hormones arriving in the blood. Androgen target cells possess an enzyme (called 5α reductase) that converts the hormones to products with about one hundred times greater effectiveness. Age, nutrition, disease, cancer, and many other factors can change the activity of the 5α reductase and also have a considerable impact on the androgen receptor that responds to the activated products. These uncertainties make the responses of cells to androgen hormones completely unpredictable.

There have been many modern attempts to stave off ageing with testosterone replacement therapy and at least one clinic in the US advertises confidently that improved health, strength, and mental well-being can be achieved by all. But these offerings remind us too readily of the animal gland extracts of earlier times and are based on a naive view of what hormones do and how they work. It is certainly doubtful whether such treatments are universally effective or safe—or even desirable.

Chapter 4
Water, salt, and blood pressure

Water

In 1995 an Essex teenager, Leah Betts, died after taking an ecstasy tablet at her 18th birthday party. Contrary to anti-drug propaganda at the time, it wasn't the ecstasy that killed her. An inquest concluded that she died of 'hyponatremia' (low blood sodium concentration), probably and tragically because she was following well-meant advice.

Ecstasy contains MDMA, an amphetamine-type chemical that produces feelings of euphoria, closeness to others, and reduced anxiety. Many of its psychological effects come from the stimulation of brain chemicals such as serotonin, adrenaline, and dopamine. Changes in these neurotransmitters are unlikely to kill you, even after a few more tablets than Leah took. To understand why she died, we need to talk about blood and urine as well as about the drug and the brain.

Dancing energetically in the close confines of a party or a club can be a hot experience. Before her party, Leah drank several litres of water as a sensible precaution against dehydration. She would have expected to lose water by sweating, which is the only way to stay cool in a warm environment, and for any excess fluid to be lost as urine.

The balance of water in the body, between what's needed and what isn't, is controlled by several hormones. One of them comes from the brain and is a close stable mate of oxytocin. It has two names. Hormone scientists call it vasopressin (actually arginine vasopressin or AVP) because one of its effects is to cause small blood vessels to contract. This reduces the space available in the circulatory system and raises blood pressure. Its other name is anti-diuretic hormone (ADH), which neatly describes its other effect of reducing urine production (diuresis) by the kidneys. We'll call it AVP (Box 5).

The extra water Leah drank would have been absorbed into her bloodstream over a short period of time. Normally, this would be detected by sensors in blood vessels of the neck and head, reacting to the increase in blood volume but also to the fall in blood salt concentration, or osmolarity. Signals from these sensors to the brain should switch AVP secretion off, allowing the kidneys to produce urine until the extra fluid volume was corrected.

Unfortunately for Leah, one of the effects of ecstasy is to *stimulate* the secretion of AVP. Because of this, she failed to make sufficient urine and her blood remained diluted by the water. The AVP probably also stopped her sweating. Excessive dilution (especially of sodium: the hyponatremia of the inquest verdict) allows water to move out of the blood and into tissues by the irresistible physical process of osmosis. This happened in Leah's brain, upsetting neural function and throwing her into a coma from which she did not recover. One could say that Leah died of water poisoning.

Vasopressin is made in the hypothalamus and secreted into the blood by the posterior part of the pituitary gland. It controls the amount of water lost in the urine by changing the way fluid moves in the millions of meandering narrow tubes, called nephrons, which make up each kidney.

Box 5. Oxytocin and vasopressin

Oxytocin and vasopressin are made in the hypothalamus of the brain and secreted into the blood by the posterior pituitary gland. They belong to a family of hormones, widespread amongst animals, called nonapeptides because they have nine amino acids. The structures below are shown as linear sequences but in reality the two cysteines (*Cys*) are linked, causing the molecule to loop. The amino acids in bold are found in all members of the family.

Vasotocin is found in all non-mammalian vertebrates (animals with backbones) and many invertebrates. Evolutionary biologists believe it may be the common ancestor of other hormones in the family.

Vasopressin and oxytocin are found in mammals. In general terms:

- Vasopressin-like hormones alter the passage of sodium and other ions across cell membranes and therefore influence the osmotic movement of water, especially in the kidneys.
- Oxytocin-like hormones cause contraction in smooth (non-skeletal) muscle cells, such as those in the mammary gland, the uterus, and the gut.

One explanation for the evolution of these separate functions in mammals, when they evolved about 150 million years ago, is the advantage of separating milk ejection from water balance and blood pressure regulation.

Vasotocin

***Cys* – Tyr** – Ile – **Gln – Asn – *Cys* – Pro** – Arg – **Gly(NH_2)**

Vasopressin (AVP)

***Cys* – Tyr** – Phe – **Gln – Asn – *Cys* – Pro** – Arg* – **Gly(NH_2)**

Oxytocin

***Cys* – Tyr** – Ile – **Gln – Asn – *Cys* – Pro** – Leu – **Gly(NH_2)**

*This is the structure in all mammals except pigs and related species: they have Lysine instead of Arginine at this position (the 8th position in the chain), making LVP instead of AVP.

Each nephron is constantly draining water and dissolved materials out of the blood to generate the basic fluid of urine. However, as it twists amongst the kidney's capillaries, the nephron's cells grab sodium ions from the drained fluid and push them back into the blood. This makes the concentration of sodium in the capillaries surrounding the nephron rather high. Driven by osmosis, water leaves the fluid and returns to the blood. (This mechanism is inhibited by caffeine, which is why you might need the toilet about half an hour after a strong cup of coffee.)

The osmotically driven water crosses the walls of the nephron by moving through cell pores called aquaporins. AVP works by making more aquaporins available, essentially making the nephron cells leakier. When AVP is there, water follows the osmotically active salt by moving through the pores into the blood, so less urine is made. When it is not there, more fluid passes down the nephron and is excreted.

The secretion of AVP varies on a minute-to-minute, hour-to-hour basis according to blood pressure and the fluid balance of the body. But it can also respond to dramatic events, for example sudden haemorrhage and shock. In the face of catastrophic blood loss, the water recovery response of the kidneys can be valuable in keeping blood pressure up and avoiding a faint. The AVP has another beneficial action: it stimulates the production of glucose by the liver. Glucose has a large osmotic effect and will assist in the reclamation of water by the nephrons, besides providing instant energy for muscles and the heart.

Another interesting situation in which AVP is secreted is during nausea. Nausea precedes vomiting and some physiologists argue that the water conserving action of AVP anticipates the fluid loss that will result from the expulsion of stomach contents. They describe it as a 'feed forward' response, using the expression in the same way as an engineer might describe a protective shut down mechanism built into a machine. But this is an uncomfortable and misleading usage because it implies that physiological systems have been designed and can somehow predict events, neither of which is true. A more acceptable explanation is that AVP happens to be secreted during nausea and the association has been conserved during evolutionary selection because it proved to be beneficial. It is of course equally possible that AVP secretion during nausea is coincidental and has nothing to do with vomiting at all.

Less dramatic but no less interesting is the observation that AVP secretion goes up at night. This reduces urine production while we are asleep, saving us the disturbance of fumbling visits to the toilet or the embarrassment of wet bed sheets. As soon as we wake up, secretion AVP goes down and normal service resumes. This day–night cycle, coupled with bladder control, is something we try to instil in children at an early age. For children who do not achieve it, and for adults for whom nocturnal enuresis is a medically significant problem, doctors can prescribe a manufactured analogue of AVP called desmopressin, sometimes offered as a nasal spray. A bedtime sniff can substantially improve the quality of life of sufferers (and their carers), provided care is taken not to pre-load with fluids.

Bedwetting can be distressing but it is a trivial ailment compared to another condition for which synthetic AVP is prescribed: diabetes insipidus. This is a nasty disease in which urine production is continuous and copious. If untreated, the uncontrolled loss of fluid generates an irresistible and overwhelming thirst, reportedly driving sufferers to desperate measures such as drinking the water from flower vases, washing-up bowls, or worse.

The insipidus epithet distinguishes it from diabetes mellitus, discussed in Chapter 6. Both conditions are characterized by large volumes of urine production. In the days before hospital labs and ingenious dipstick tests, the only way doctors could tell them apart was by tasting the urine: one urine would be tasteless and the other sweet. The medical profession must surely be grateful for advancements in clinical science.

Diabetes insipidus happens either when the brain/pituitary system fails to secrete AVP or when the AVP no longer works on the kidney nephrons. The former could be the result of a tumour, a head injury, or a side effect of surgery, and it can be treated easily with the synthetic AVP. The latter may be caused by kidney disease or the autoimmune production of antibodies against nephron AVP receptors, and is rather more difficult to treat.

Salt and blood pressure

An important context for understanding how AVP regulates water in the body is blood pressure control. Doctors measure blood pressure because it can be a direct indicator of health, a sign of existing or incipient disease, and a way of monitoring aspects of lifestyle. In terms of simple physics, blood pressure goes up or down according to three things: the volume of the blood, the space available for it in the circulatory system, and the force applied to it by the beating heart. These three things fluctuate all the time, reflecting both sudden changes (posture, activity, emotion, fluid balance) and more gradual changes (lifecycles, lifestyles, disease, stress).

As we saw, the reclamation of water by kidney nephrons, operating through osmosis, is a reaction to the movement of sodium ions. Blood volume and pressure are also adjusted by changing the amount of sodium ions reclaimed by the nephrons. Blood filtration happens at the top end of each kidney nephron, in a structure called the glomerulus. Fine capillaries bring blood from the body into each glomerulus; other capillaries carry the depleted blood away, but these are wrapped around later sections of the nephron and reclaim much of what was lost. The walls of the capillaries entering the glomerulus are made of stretch-sensitive cells, called J-G cells, which constantly monitor the pressure of the incoming blood.

If the pressure goes down, the J-G cells secrete a hormone called renin into the general circulation. Renin is really an enzyme and its job is to start the activation of a protein in the blood. (Renin is not to be confused with rennin, the milk-curdling enzyme in the stomach.) It shortens this blood protein and makes it available to a second enzyme that is waiting in the blood, called ACE. ACE shortens the protein even further, turning it into a hormone called angiotensin II (or Ang-II for short).

Ang-II does two things, both of which raise the blood pressure. Its first effect is to make capillaries around the body contract. This reduces the space available for the blood, increases the tension in blood vessel walls, and raises the pressure. Its second effect is to stimulate cells in the outermost part of the adrenal gland to secrete a steroid hormone called aldosterone. Aldosterone acts back on the kidney. It stimulates the mechanism in nephron walls that reclaims sodium ions from the forming urine. Water follows the sodium out of the urine by osmosis and the volume of the blood goes up.

Hormones from the heart

Working together, Ang-II, aldosterone, and AVP do a pretty good job of balancing blood volume and circulatory space to keep pressure stable when the volume and dilution of the blood change. But what happens if the concentration of salt (sodium and other ions) in the blood starts to rise, for example after a salty meal or from excessive sweating? Is there a direct way to get rid of excess salt?

In 1981, Adolfo de Bold and his colleagues in Toronto asked themselves exactly that question. Under the microscope, they had noticed that cells in the atria of the heart, the upper chambers that receive blood from the body and lungs, contained granular structures alongside the machinery needed for contraction. These granules were rather like those in hormone secreting cells in other organs and they wondered what they were for.

They killed some rats, made extracts of their atrial tissue and prepared to test them. Fortunately, ventricular tissue has no granules, so extracts of that made a convenient control for their experiments. They injected the extracts into the veins of anaesthetized rats and watched what happened to the urine. Within two minutes, the rats' urine production started to increase. Over the next 20 minutes or so, the amount of salt in the urine went up 30 times. Even though more urine was being produced, the rate of blood filtration by the kidneys did not change. When they examined samples of blood they found that the number of red cells in each millilitre had gone up, clearly showing that blood volume had been lost. They also checked the cardiovascular system and found that blood pressure had dropped while the heart rate stayed the same.

De Bold's team could not explain these effects by any of the hormones already known to regulate urine production and blood pressure. They concluded that the heart atria must be producing something that reduced the movement of sodium over the nephron wall and increased its loss in the urine. The effect of this was to reduce the water uptake by osmosis, lower the blood volume, and increase the amount of urine produced.

They eventually purified from the heart extract a small protein, made of 28 amino acids, and called it atrial natriuretic (meaning sodium excreting) peptide or ANP. Secretion of ANP turned out to depend on the stretch experienced by the atrial muscle cells. The amount of stretching represents the pressure in the cardiovascular system and will increase if blood volume, such as that caused by increased salt and osmosis, goes up.

Not only does ANP reduce sodium reclamation by the nephron, it also inhibits the production of aldosterone by the adrenal gland, renin by the kidney, and vasopressin by the brain. It shows rather nicely how the systems that adjust blood and urine volume are closely integrated, accounting for the wonderfully fine and sensitive way in which blood volume and pressure are regulated.

Since de Bold's discovery, other types of ANP have been discovered. They are produced in other parts of the body including the brain and by the placenta during pregnancy. Some of them affect the way the heart works, changing the frequency of beats and the amount of blood pushed out by each contraction. These characteristics of the heart, which are also controlled by nerves and by adrenaline, provide further ways of adjusting the pressure of the blood.

The renin–Ang-II–aldosterone–ANP system has enough complexity and logic to make it a favourite target for blood pressure researchers. Recent research suggests that tiny, previously unrecognized tumours in the adrenal gland may be a cause of hypertension. The aldosterone, which causes the kidneys to retain salt, normally comes exclusively from the outer layer of the adrenal cortex, called the zona glomerulosa or Z.g. for short. Highly sensitive scans carried out in young adults can sometimes pick up benign cancer cells with the characteristics of Z.g. cells growing in one of the other layers of the cortex called the zona fasciculata (Z.f.). The Z.f. normally secretes cortisol rather than aldosterone (Box 5), so this unexpected discovery has the potential to explain incipient hypertension that may become a problem in later life.

Inhibitors of ACE, the blood protein that activates Ang-II, are amongst the most commonly prescribed blood pressure lowering drugs. Other drugs block Ang-II receptors on blood capillaries, stopping them from contracting. These also help to block the action of a contraction-stimulating hormone produced by the capillaries themselves called endothelin. Endothelin inhibits the sodium- and water-reclaiming system in the nephron, rather like ANP, so perhaps this effect could be a therapeutic target too.

Chapter 5
The hard stuff: calcium, cells, bones, and cancer

Calcium all around us

The crumbling coastal cliffs that dominate the beaches in many parts of the UK are great places to find fossils. It is easy to collect beautiful fossil shells of myriad shapes and sizes. With care, a little perseverance, and a lot of luck, one may even find the bones of extinct animals.

These attractive and intriguing remnants remind us that a large proportion of the limestone on our planet is made from ancient biology. Throughout the history of life on Earth, single-celled and multi-celled organisms have used calcium and its minerals to make protective and supportive structures. We see the resulting sediments in mountains, rocks, caves, and quarries.

It is hardly surprising that living organisms have exploited calcium in this ubiquitous way. Calcium is the fifth most abundant element in the earth's crust (after oxygen, silicon, aluminium, and iron). Most chemicals containing calcium, including salts like carbonates and phosphates, are poorly soluble in water, and settle out very easily. This quirk of calcium chemistry makes it perfect for building hard materials that defy the pull of gravity and protect against predators. It also allows for continuous recycling and reconstruction.

But that same chemistry also makes calcium potentially dangerous to life. Precipitating salts would make a complete mess of biochemical and biophysical systems if they appeared in the wrong place or at the wrong moment. Many essential life processes, from the contraction of muscles and transmission of nerve impulses to the fertilization of an egg by a sperm and the production of enzymes to break down food, depend on calcium. They use the element in its soluble, highly reactive, positively charged (cation) form, written as Ca^{2+}. Not surprisingly, animals have evolved highly sensitive hormone systems to keep the correct balance between these reactive ions and the precipitating salts.

Let us look first inside a typical cell. Cells need calcium for all sorts of processes including expression of genes, use of energy supplies, transmission of electrical impulses, conversion of one molecule into another, change of shape, movement, and division. The trouble is that their cytoplasm also contains phosphate ions. These are being constantly shuttled from one molecule to another, perhaps in response to an outside stimulus or during the metabolism of important proteins and complex sugars. Phosphate ions are negatively charged (anions; PO_4^-) and, like calcium ions, are highly reactive.

If calcium and phosphate ions meet each other at anything above a miniscule concentration, they form crystals and settle out. Other anions such as carbonate and sulphate react with calcium in a similar way, although there is rather less of these inside most cells. Cells avoid the precipitation problem by hiding their calcium ions in a special compartment called the endoplasmic reticulum. They squeeze the calcium ions out into the cytoplasm in tiny amounts, only when needed and carefully protected by protein.

The same risk of precipitation arises outside cells—in tissue fluids and blood. The hard materials of bone and shell, made of calcium phosphate and calcium carbonate, respectively, develop in places where the ion concentration has been allowed to rise. Everywhere else, hormones ensure that concentrations stay within safe limits. The clever thing is that these hormones also ensure that enough calcium remains available for cells to function properly.

Keeping calcium under control

Mammals give their offspring a developmental head-start by passing calcium from the mother in the form of milk. The dairy cow, which has been carefully bred for decades to produce lavish quantities of milk, illustrates the problem of getting calcium ions in the right place at the right time. A cow's udder starts to fill at the end of pregnancy, enabling the calf to suckle shortly after it is born. But there is a limit to how much calcium a cow can absorb from food, even if the stockman deliberately adds it to the diet, and as birth approaches the udders manufacture milk at a rate that outstrips the supply of calcium.

The level of calcium in the cow's blood starts to fall and hormones respond immediately to mobilize calcium stored in the skeleton. But even this may not be enough and the animal's blood calcium can continue to sink rapidly. This upsets the transmission of impulses by nerve and muscle cells. In many cases the effect is sub-clinical (not obvious) but it can stop organs like the gut and uterus from functioning efficiently and may make the animal more susceptible to disease. In the worst cases, the animal loses control of its muscles, begins to shake, and may eventually fall over and become unconscious. Fortunately for herdsmen and their precious stock there is a simple remedy: a rapid infusion of a calcium-rich liquid directly into the cow's blood will get her on her feet and back to full health in remarkably short order.

The hormone that normally keeps blood calcium up at the right level is called parathyroid hormone or PTH. This small protein is secreted by a pair of tiny glands next to the thyroid gland in the neck. Parathyroid gland cells are, in effect, highly sensitive calcium ion sensors: they jump quickly into action when the concentration falls below a critical level, and they switch off just as fast when the level is restored.

PTH does several things. On the one hand it conserves calcium by reducing the amount that the kidneys let out into the urine. At the same time, it causes bone cells to dissolve some of the calcium phosphate of which the bone mineral is made. Normally, the release of just a tiny amount of bone is enough to bring the blood calcium level back to normal, so the strength of the skeleton is unaffected. Phosphate from the bone is released into the blood along with the calcium, and PTH makes the kidneys excrete more of this into the urine too.

Bone is a remarkable material. It is easy to be misled by its solidity, and also by its persistence in the environment long after bodies have decayed, into imagining it as stable and immutable. In fact, bone is constantly recycled and reworked. In adults, around 10 per cent of the material in the skeleton is turned over in the course of a year. This keeps it mechanically strong and ensures that it changes shape in response to the stresses and strains of body movement. The rate of turnover is about ten times higher in children as they grow, and in the earliest few years of life the whole of the skeleton is reformed several times over.

Mixed into the bone structure are a large number of cells. Some of these bring calcium and phosphate ions together so that they precipitate into the solid material. Others—the ones that respond to PTH—have the opposite effect. They make the local conditions very slightly acid so that the calcium phosphate is very slightly more soluble. (The delicacy of the chemistry involved in this precarious balance is astonishing.) Coupled with the effects of PTH on the kidney, this system keeps the blood calcium ion concentration within very tight limits.

Taking it in

The ins and outs of bone make up one end of the calcium balance system. At the other is the absorption by the gut of calcium delivered in the food. This happens mostly in the first part of the small intestine, in a region called the duodenum, and is helped by the acidification of food materials within the stomach.

Calcium can only cross the wall of the duodenum and move into the blood with the assistance of a molecular shuttle. This is a protein called calbindin, which, as its name suggests, binds calcium. This is where a second calcium regulating hormone—vitamin D—enters the picture: the cells of the duodenum will only manufacture calbindin when the gene for it is switched on by vitamin D.

Readers of the author's generation will recall this vitamin being given to them as children in daily spoonfuls of cod liver oil. (Slightly younger readers may have been more fortunate to receive shiny yellow, easy-to-swallow capsules.) In those dark but optimistic post-war days, vitamins were the new kids on the preventive medicine block. With a limited and unreliable range of foods available there was wisdom in preventing deficiencies of these essential co-nutrients, and dietary supplements certainly did us no harm. Coupled with the calcium supplied by free school milk, we probably grew up with the strongest bones in world.

What was not really understood at the time was how vitamin D actually works. It is clear now that it has to be absorbed from food, activated and re-routed back to the intestine in order to switch on the calbindin gene. But it does a large number of other things too, not all of them directly connected with calcium. Because of these multiple effects and because of the involvement of internal organs in its production, vitamin D is now more properly called a hormone than a vitamin.

The vitamin D in fish oils and similar food products is a complex molecule related to cholesterol, called calciferol (or vitamin D_2). A very similar material, called cholecalciferol (or vitamin D_3), is made in the skin from a modified form of cholesterol (Box 6). To make it, some complex chemical bonds need to be smashed and only the ultraviolet wavelengths of the light spectrum have enough energy to do this. This is why exposing the skin to sunlight is such a good way of preventing vitamin D deficiency.

Once cholecalciferol has been made in the skin or absorbed from food, it circulates in the blood. But as yet it is inactive. First the liver, and then the kidneys change its structure into (something with a tediously long name abbreviated to) 1,25DHCC. This is the active chemical that causes calbindin to be made in the small intestine.

But 1,25DHCC has many other effects. In particular, it increases the activity of both types of bone cell—the ones that bring calcium and phosphate together and the ones that cause bone breakdown. This two-directional coupling may seem counter-intuitive but it means that the bone becomes more active and its mineral is recycled at an increased rate. This has several advantages. In growing bones it means that the existing mineral can be combined with new material and deposited in a longer, wider, or thicker way. In both growing and fully grown skeletons it means that bone can be strengthened in response to the pull of active muscles. The well-known bone weakness of astronauts in weightless orbit results from the continuing turnover of bone mineral in the absence of these stresses.

As we saw with the dairy cow, when there is insufficient calcium in the diet or when demand is especially high, PTH and 1,25DHCC normally combine to release some of the calcium stored in bone into the blood. For people on a reasonably balanced diet, calcium deficiency leading to bone weakness is unlikely if there is an adequate supply of vitamin D. Yet some doctors believe that

Box 6. How vitamin D is made

Vitamin D is a steroid hormone, made from cholesterol, in which one of the ring structures has been broken open by ultraviolet light. This happens in the skin of animals, as shown below, but a similar reaction also happens in the leaves of some plants, making them a valuable source of the vitamin. Industrial production of vitamin D involves ultraviolet treatment of 7-dehydrocholesterol derived from lanolin.

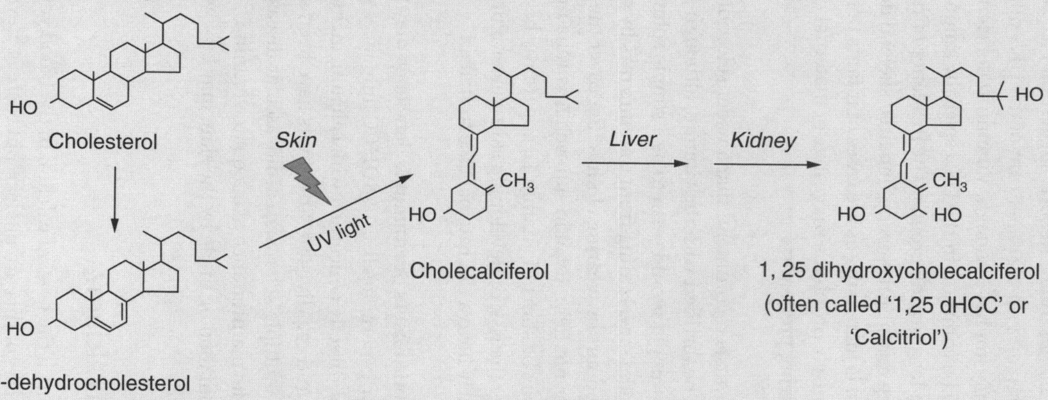

HO

Cholesterol

Skin

UV light

HO

7-dehydrocholesterol

CH_3

HO

Cholecalciferol

Liver

Kidney

HO

CH_3

HO

HO

1, 25 dihydroxycholecalciferol
(often called '1,25 dHCC' or
'Calcitriol')

Cholecalciferol (Vitamin D_3) and the plant equivalent (ergocalciferol, vitamin D_2) are inactive. Their conversion into an active hormone (1, 25 dHCC) needs enzymes in the liver and kidney.

These chemicals and others in the vitamin D system are known by several other names.

As with most hormones, well meaning attempts to unify the nomenclature have increased rather than reduced confusion.

calcium deficiency, far from being a problem of the past, is common in Western societies today. Reports suggest that a quarter of British children and up to 90 per cent of the UK's multi-ethnic population may not have enough. Vitamin D deficiency in adults is reported to be common in the US, especially amongst poorly educated and low income groups. The incidence of rickets, a bone-thinning condition causing impaired skeletal development and fractures, is increasing as are cases of infants born with heart, nerve, and muscle problems suggestive of maternal calcium deficiency during pregnancy.

Some nutritionists recommend that a wider selection of processed foods should be fortified with the vitamin, although greater exposure to sunlight would seem to be a simple solution to the risk of calcium deficiency. Hiding from the sun's rays by spending too much time indoors or wearing clothes that cover the whole body, certainly does not help. Even in sun-rich Australia up to a third of under-25s may be short of vitamin D, apparently because of anxieties over the risk of developing skin cancer. Public health advisors clearly have a delicate balance to strike.

But more subtle effects of vitamin D deficiency may be equally important for general health: 1,25DHCC influences several processes not directly concerned with calcium, including immune competence and cell division. Low levels have been associated with cancer, multiple sclerosis, cardiovascular disease, and asthma. So the maintenance of adequate vitamin D, whether absorbed from food or made by the skin, may be a wise policy more generally.

Too much calcium?

What happens if the level of calcium in the blood gets too high? The health risks of this are unclear, but there is, nevertheless, a hormonal response to the situation. The tale can be traced back to a salmon processing factory in the southwest of Canada.

In the early 1960s, Douglas Harold Copp (1915–98), a Canadian biochemist who had done secret work on radiation and bone marrow for the Allies during World War II, set up a hormone research group at the University of British Columbia. It was known that surgical removal of thyroid glands from dogs upsets their calcium levels, and Copp's team set about identifying the crucial secretions that normally maintained homeostasis.

They experimented by flushing the thyroid organs of dogs with fluids both rich and poor in calcium, and in 1962 discovered a protein secreted in response to the high calcium solution. When purified and re-injected, in classic Doisy style, it lowered the blood calcium level—exactly the opposite effect to PTH. It was named Calcitonin. The thyroid and parathyroid glands are hard to separate and it was originally thought that the parathyroids were the source of both hormones, but further studies showed that special cells within the thyroid, called C-cells, were the source of calcitonin.

After calcitonin had been identified in the thyroids of several other mammals, it was realized that the C-cells of non-mammals (birds, amphibians, and fish) occurred in separate tissues called the ultimobranchial glands. The highest amounts were found in salmon, with Copp famously reporting:

> In 1968 we arranged with the Canadian Fishing Company of Vancouver to collect 200lbs [90kg] of ultimobranchial tissue from approximately 5000 tons of [British Columbia] salmon. From this material... pure salmon calcitonin was isolated.

Despite their activity, the ultimobranchial glands are tiny and one has to admire the fortitude of Copp's wife faced with his return from the factory at the end of each working day. But such were the sacrifices of endocrine pioneers in the heyday of hormone discovery.

Salmon calcitonin is a great deal more potent (has more calcium lowering action for a given amount) than mammal calcitonin, and it became the standard preparation for research and clinical use. Salmonids and other migratory fish need so much calcitonin because they have to cope with changing calcium levels in the water as they move between seas and rivers. Salmon calcitonin has long been used to treat bone conditions such as Paget's disease, although it is now made synthetically rather than by extraction from fish.

Calcitonin encourages bone cells to deposit calcium in the skeleton and causes the kidneys to excrete calcium in the urine. Thus the ups and downs of blood calcium can be regulated by a simple balance between calcitonin and PTH. However, the importance of calcitonin in humans is the subject of controversy. Hypercalcaemia (clinically raised calcium levels) is rather rare and there are no significant diseases from calcitonin deficiency or over-secretion; so perhaps the hormone is 'vestigial'—a useless evolutionary remnant. Against this, the logic of evolution by natural selection says that if a hormone exists it must have a function (Chapter 9 and Box 10 in that chapter). Calcitonin protects the skeleton during lactation and does a few other things too, such as controlling appetite, so perhaps its name and the experiments showing its dramatic effect on blood calcium mislead us about its most important functions.

A more informative view of calcium and its regulating hormones comes by reconsidering bone itself. Bone formation and recycling take energy, as well as calcium and phosphorus, and all aspects of skeletal development are affected by the hormones of growth and metabolism (growth hormone (GH), insulin-like growth factor I (IGF-1), thyroid hormone, and cortisol, to name just a few). The well-known loss of bone mineral after the menopause is due to low levels of oestrogen and is a reminder of the importance of reproductive hormones in skeletal development. The brain seems to be able to monitor the status of the skeleton, and there are connections between bone development and appetite that work through serotonin, leptin, insulin, and hormones produced by the gut, in addition to calcitonin.

An exciting recent discovery is that bone itself is a hormone-secreting tissue. The bone cells that join calcium and phosphorus together produce a small protein called osteocalcin. This was once thought to be exclusively concerned with the local mineralization process, but it is now known to be secreted into the blood. Osteocalcin stimulates insulin secretion from the pancreas (providing another link between bone and energy) and stimulates testosterone production by the testis (providing another link between bone and reproduction). Bone is far from just an anti-gravity device.

Chapter 6

Appetite, fat, and obesity

Insulin and sugar

On 6 December 2008, an American heiress and socialite Martha (Sunny) von Bülow, died aged 76 after being in a deep coma for 28 years. She collapsed first in December 1979 and was revived in hospital, but she collapsed again the following Christmas, this time never to recover. Medical experts attributed her brain injury to severe hypoglycaemia (very low blood sugar).

Sunny's husband Claus von Bülow was tried twice for her attempted murder. At his first trial in 1982, metabolism expert Dr George Cahill testified that Sunny had been injected with insulin. Insulin causes blood sugar levels to fall, and it was known that a large injection would cause prolonged suppression. An insulin-contaminated syringe was presented in evidence. Von Bülow was convicted and sentenced to 30 years in prison, but this was overturned on appeal.

At von Bülow's second trial in 1985, defence experts convinced the jury that Sunny's death was due to ingested rather than injected drugs, including a very large number of self-administered aspirin tablets. Doubt was cast on Cahill's evidence, including the syringe, and he subsequently admitted that he could not be sure of his earlier conclusion.

The von Bülow story became a book and a film (*Reversal of Fortune*), undoubtedly contributing to the rare fame, even notoriety, experienced by a humble hormone. But poor George Cahill deserves a better reputation than that of an unreliable witness. As a researcher, he had investigated the metabolic effects of starvation diets. These had been proposed as a radical treatment for extreme obesity. He monitored insulin in the blood but also looked at how the body's protein and fat were broken down to supply energy to the brain when energy from food was in short supply. He found that a fall in insulin concentration allows those other energy sources to become available. In particular, fat stores can be turned into ketones, which the brain can use in place of glucose.

Insulin and fat

Fat is extremely 'energy dense'. That is to say, any particular amount contains much more energy (9 Calories—or kilocalories—per gram (Cal/g)) than would be held in the same weight of protein (7 Cal/g), complex carbohydrate, or even pure sugar (4 Cal/g). This explains why it evolved as the best way to carry reserve energy in the body. Some believe it also explains why humans have evolved to find the flavour and texture of fatty foods (for example butter, cream, and other animal products) so appealing. Eating fat is the most efficient way to consume energy, and storing it in the body provides a buffer against the ups and downs of food availability. But as we know, it is easy for the storage to get out of hand especially when food is cheap, abundant, convenient, and tasty.

Insulin stimulates this storage process and it is the most anabolic or tissue building of all hormones. Fat cells respond to insulin by taking glucose from the blood and converting it into lipids. Muscle cells take up glucose and make protein. The liver responds by turning glucose into glycogen, which is a way of storing energy as carbohydrate.

These processes combine to reduce the amount of sugar in the blood and they start within minutes of eating a meal. Making the fat and protein takes a little longer. The whole set of responses can vary in intensity over days and weeks according to the amount of food consumed, the amount of exercise taken, and numerous other variables of daily life.

Research on insulin and its actions reveals a network of frightening complexity, involving not just fat, muscle, and liver, but brain, gut, heart, kidney, and much else besides. In recent times, it has been possible to study mice in which particular genes have been inactivated, either throughout the animal or in single tissues. The silenced genes could be for insulin, insulin receptors, or individual cogs in the biochemical gearbox inside cells. It is possible to see what happens when the enzymes that make fat or glycogen are inactivated, or when the molecules that move glucose across cell membranes are absent. Typically, as soon as one system is shut down, another is revealed. The question frequently shifts from 'What does this bit do?' or 'How does this change that?' to 'Is there anything that hasn't been affected?'.

Insulin and affluence

The need to get to grips with this complex system is unavoidable from medical, social, and economic points of view. In 2012, doctors in England wrote out 40 million prescriptions for the treatment of diabetes mellitus. They also performed 100 diabetes-related foot amputations. About 2.5 million people have the disease and the number is rising. Worldwide, the number of sufferers approaches 350 million. About 90 per cent have Type 2 diabetes, the prevalence of which increases sharply in societies where people get richer, eat more sugary and fatty foods, and take less exercise.

Diabetes mellitus comes in several forms, some more common than others. In all of them, as the sugar level in the blood rises uncontrollably, the kidneys cannot stop it leaking into the urine. The osmotic or water-drawing effects of the sugar mean that more urine is produced and the sufferer experiences a corresponding thirst. The symptoms explain the name. 'Diabetes' means the 'running through' of copious urine. 'Mellitus' means sweet, and describes the urine's taste (as a traditional but thankfully no longer recommended way of distinguishing it from the quite different disease called diabetes insipidus, as discussed in Chapter 4).

Some rare forms of diabetes are inherited or congenital, and another type develops in pregnant women. Type 1 diabetes, which was previously called 'insulin dependent' or 'juvenile onset', is an autoimmune disease in which a person's immune system attacks the insulin-secreting β-cells of the pancreas. The cells are eventually destroyed to a level where insulin production is inadequate and a medical emergency arises. Treatments that try to suppress the immune system have only limited effect, making replacement of the missing insulin the only currently effective option.

This usually involves the patient monitoring their blood sugar, by regular pin pricks or saliva testing, and injecting an appropriate amount of hormone. It is an inexact, uncomfortable, and unremitting procedure and each year there are a number of deaths when things going wrong. Most patients get the hang of it even if it does interfere with everyday life. In the future, implanted systems for automatic sugar monitoring and insulin delivery, almost amounting to a replacement pancreas, should improve the lives of these people dramatically.

Type 2 diabetes used to be called 'maturity onset' to distinguish it from Type 1 and perhaps to reflect its position as a so-called lifestyle disease. Some estimates suggest that more than a quarter of over 65s eventually develop it. Well established risk factors include obesity, especially excess fat around the abdomen, and a combination of symptoms called metabolic syndrome. These symptoms include high blood pressure, low blood HDL ('good' cholesterol), high blood triglyceride (the building blocks of fat), and high resting blood glucose. They appear before the diabetes itself gets going, which is why health checks can be the key to prevention.

But what exactly is the connection between obesity and poorly regulated blood glucose? The link we are looking for here is between excessive fat stores and a combination of impaired insulin secretion and resistance to insulin action. A current explanation is that obesity is a kind of inflammatory condition in which the fat cells increase in number and produce hormones called cytokines. The normal role of cytokines would be to coordinate defences against impending infection, but instead they set off some unwanted reactions. In particular, they damage the insulin-secreting β cells of the pancreas and inactivate insulin receptors. They probably also contribute to the hypertension and cholesterol-related symptoms.

The damage to insulin receptors in muscle, liver, and fat cells means that sugar cannot be removed in sufficient quantity from the blood. It also means, of course, that insulin becomes a less and less effective method of treatment, leading to an earlier description of the disease as 'non-insulin dependent'.

Fortunately, the cells of many important organs, including the brain, kidneys, liver, heart, and gut, can use blood sugar without the action of insulin. This means that their own energy supply is unaffected, even though they may be seriously damaged in other ways. The muscles of the skeleton can use glucose independently of insulin so long as they are active, and this is why gentle exercise can help to reduce the blood sugar level in a steady manner. Doctors and dieticians also advise diabetics to eat foods with a low glycaemic index—those that release their carbohydrate slowly during digestion. Eating these will not cure the disease but they will attenuate the sugar hit, thereby requiring less of an insulin response.

Experiments in animals seem to show that insulin and diet affect lifespan. Worms and fruit flies can be made to live for 50 per cent longer than normal by putting them on low energy diets. Mice will live about a year longer and they also suffer from fewer life-threatening diseases. These effects are related to lower levels of insulin and some researchers think that it is the reduced effect of the hormone on the brain that is the most important effect. Sorting out this issue in humans and other mammals is tricky because we have longer lifespans, and many other variables, including genetic ones, influence the age at which we die. Nevertheless, some gerontologists extrapolate from animal models and advise us to reduce calorie intake if we want to live longer.

Sugar and survival

The legacy of evolution and the struggle for survival—which is assumed to have entailed a relentless search for food—means that the body has many more hormones capable of responding to energy shortage than to energy abundance. Some describe this as metabolic 'thriftiness', although the word tells us nothing about mechanisms. The physiological fact is that when food is scarce and insulin is out of the way, other hormones including glucagon, adrenaline, vasopressin, growth hormone, and cortisol allow the body to make use of its stored energy.

Under normal circumstances, the amount of glucose that circulates is rather limited: a man of average height has in his blood the equivalent of about 125 grams of pure sugar. This quantity wouldn't last very long, especially if the muscles that work the skeleton are active. The α cells in the pancreas are constantly monitoring the blood sugar concentration and as soon as it starts to fall they secrete glucagon.

Glucagon's main role is to release the energy stored as glycogen in the liver. The system is so sensitive that in normal, adequately fed individuals blood glucose concentrations seldom fall. The rate of glucose flux—its throughput from liver to blood to organs (muscle, brain, kidney, and everything else)—increases, but the blood concentration remains impressively stable.

The liver holds about 75 times as much glucose as is circulating in the blood. A third of this is readily available as glycogen and another third is slightly less readily available as fat. Normally, the glycogen store is replenished from food, stimulated by insulin, but if this supply is inadequate other systems come into play. Growth hormone and cortisol stimulate the liver to make glucose from fat, and the liver can also use lactic acid, which is a by-product of intense muscle activity. Because fat is a rich energy source—1 gram is equivalent to about 2.4 grams of sugar—it doesn't take much to restock the liver and keep blood levels in check. But what happens if times are so tough that the fat runs out?

In 1992, Ranulph Fiennes and Mike Stroud set off on their epic attempt to make the first unsupported crossing of the Antarctic. According to their calculations, they expected to use energy at the rate of 27 million Joules (MJ) or 6,500 kilocalories per day. That is roughly the equivalent of running two marathons every day.

Noting the energy density of fat, Fiennes and Stroud realized that the only practical way of taking sufficient food with them would be to live almost entirely on butter for the duration of the 2,250 kilometre (1,350 mile) journey. Despite this unappetising prospect, they still expected to be about 15 per cent short of the energy needed.

In the event, their actual energy expenditure over the 95 day journey turned out to be nearer 30 MJ per day. This represented a daily shortfall of 9 MJ, equivalent to about seven Mars bars or half a bag of granulated sugar. Not surprisingly, despite fattening themselves up before they set off, they each lost about a third of their body weight.

The hormones of starvation (for that in effect is what Fiennes and Stroud imposed on themselves) are growth hormone (GH) and cortisol, and these would have released body fat to keep energy supplied to the muscles. They would also, as Cahill discovered, have provided ketones to keep the brain functioning. But when the body fat ran out, the only useful energy source remaining would be the protein of the muscles. Under these circumstances, GH and cortisol cause the liver to convert that protein into glucose. Muscle wastage, especially in muscle groups not being used very much, is the inevitable result.

Fiennes and Stroud survived their remarkable journey, but perhaps one of the most interesting things about the experience was what happened after they returned. They describe an uncontrollable and almost continuous hunger while their bodies put back the lost protein and fat. Yet the moment they regained their normal body weights, the hunger stopped.

Appetite control

As well as demonstrating the remarkable ability of the body to recover from energy exhaustion, this story illustrates perfectly the close association between the energy requirements of the body and appetite. It is a relationship that has intrigued researchers for many years yet, whilst we know that hormones are the key, it is a complex physiological system that remains poorly understood.

The fat, or white adipose tissue, which Fiennes and Stroud replenished on their return secretes an appetite-regulating hormone called leptin. This small protein inhibits the centres in the brain that drive the search for food. The more fat cells there are, the more leptin is produced and the lower the appetite.

Well, that is the theory, and the effect can be demonstrated quite easily in laboratory rats, but it is not really that simple. For one thing, leptin doesn't always work: in some individuals the brain's appetite centres have missing receptors and the hormone might as well not be there. There is also a rare condition in which individuals have a genetic inability to make the hormone itself. But even in the vast majority who have neither deficiency, it is easy for leptin's effects to be overridden.

Furthermore, we mislead ourselves if we think that appetite is a gently undulating drive, tied to a cycle of three square meals a day, and subtly adjusting itself to our energy needs. Reflect for a moment on the close association between the desire to eat and other emotions and circumstances (anxiety, excitement, anticipation, boredom, depression, exercise, social participation, family cohesion, etc.), as well as the irresistible attractiveness of certain food odours and our revulsion to items we dislike or suspect to be contaminated. There has to be more to appetite than automatic hormonal responses.

When the function of insulin was first discovered, it was thought that the fall in appetite we experience after eating might be easily explained by its rise in concentration as food is absorbed. A refinement of this idea suggested that the brain was responsive to glucose as well as insulin and that hunger occurred when both signals were absent. Animal experiments showed that cells in parts of the hypothalamus where appetite seemed to be regulated were indeed directly glucose- and insulin-sensitive.

The problem was that the appetite centres were also responsive to many other things, and other areas not associated with appetite could be influenced too. The whole concept became very messy. It also suffered from being grounded in the simplistic and now out-dated view that the brain has separate regions controlling individually identifiable physiological responses.

A further complication is that different timescales are involved. Leptin may adjust appetite over a period of days or weeks, but that probably has little to do with the feelings of hunger or satiety associated with individual meals. Similarly, while the insulin response to food may reduce immediate hunger, the general level of insulin in the blood, like that of leptin, reflects the amount of fat in the body. So the two hormones probably cooperate in adjusting long-term energy balance.

We now know that a very large number of hormones can affect appetite. Some interesting ones (together with their confusing but commonly used abbreviations) are shown in Box 7. They are secreted into the blood from cells in various parts of the gut. Nearly all of them are made in the brain as well. Some have related structures and fall into family groups.

Box 8 shows how the amounts of three of these hormones—ghrelin, insulin, and leptin—change in the blood in relation to meals. Ghrelin seems to rise in anticipation of food, perhaps as an indication of hunger, while insulin rises after meals, reflecting the need to keep blood glucose under control. Leptin's profile barely fluctuates around meal times, being dominated by a diurnal rhythm rather than short-term changes.

With this complexity it is hardly surprising that appetite, despite its importance for survival and health, is resistant to simple physiological dissection. These hormones may provide mechanisms but they are not necessarily the origin of behaviour or disease. The biggest difficulty is that they perform against the vast psychological and emotional backdrop provided by the conscious and subconscious mind, located in the so-called higher centres of the brain. We also need to factor in a network of entrained rhythms associated with daily, seasonal, annual, and reproductive cycles. Stages of growth, age, and health provide further influences.

Box 7.

Some of the hormones produced by the alimentary tract and pancreas act on the hypothalamus and other parts of the brain to alter appetite. PYY and PP are related to a brain chemical called neuropeptide Y (NPY) and work through similar receptors. GLP-1 and OXM are from the same family as the blood sugar lowering hormone glucagon. Leptin is not secreted by the gut but by white adipose tissues (fat); it can have a long-term effect on appetite.

Hormone	Main source	When secreted	Effect
Cholecystokinin (CCK)	Small intestine	Immediately after a meal	Briefly reduces appetite Adjusts intestinal mobility and the passage of food Increases brain sensitivity to leptin
Peptide YY (PYY) Pancreatic polypeptide (PP)	Large and small intestines Pancreas (Gamma cells in the Islets of Langerhans)	After a meal and as food passes along the gut	Strongly reduces appetite Concentrations tend to be lower in obese people and higher in anorexics
Glucagon-like peptide-1 (GLP-1) Oxyntomodulin (OXM)	Small intestine (ileum) Large intestine (colon)	In response to food (at the same time as PYY), in proportion to calorie content	Helps to suppress appetite Adjusts the amount of insulin secreted by the pancreas
Ghrelin	Stomach	During hunger (falls soon after eating) Has a diurnal pattern similar to that of leptin	Increases appetite (the only hormone known to do so) Concentrations tend to be higher in obese people and lower in anorexics

Appetite-controlling medicines?

If there were straightforward hormonal explanations for appetite, Type 2 diabetes, metabolic syndrome, obesity, and other equally troubling appetite-related conditions such as anorexia and bulimia, they would have been found by now. It is safe to say that there are not. The pharmaceutical industry, which is always keen to make money out of magic pills for common ills, has invested vast sums into appetite and obesity research. So far, their shareholders have received rather meagre returns.

Some recent pharmaceutical research illustrates the difficulty of translating basic hormone science into useful medicines. The tiny peptide hormone somatostatin is produced in the brain, where it suppresses GH secretion; in the pancreas, where it moderates insulin and glucagon secretion; and in the stomach and bowel, where it slows the progress of digestion. With its GH-regulating hat on, somatostatin reduces the rate of metabolism. Logic suggests that it would, as a result, tend to increase fat deposition and body weight, and this is indeed what it does.

Experiments in mice showed that somatostatin's effects could be counteracted by tricking the immune system into producing somatostatin-neutralizing antibodies. This was done by injecting the animals with a modified form of the hormone and causing an immune response, in much the same way as an attenuated virus or bacterium is used to induce immunity to disease. The treated animals rapidly lost weight, even when fed a high fat diet, and they did so without reducing the amount of food they ate.

Do these experiments hold out the prospect of a 'flab jab' as an effortless, non-surgical remedy for the dire consequences of overeating? It seems unlikely. Fundamentally, obese humans may not respond like mice (and ethical permission for the necessary experiment might be hard to obtain). A treatment that reduces body weight without affecting appetite must, by the laws of

Box 8. Hormones and meals

Blood profiles of Ghrelin, Insulin, and Leptin in 10 human subjects over 24 hours. Normal meals (B: breakfast, L: lunch, D: dinner) were eaten at the times indicated.

The concentration ranges of the hormones are Ghrelin 300–700 pg/ml, Insulin 100–500 pmol/ml, Leptin 15–25 ng/ml.

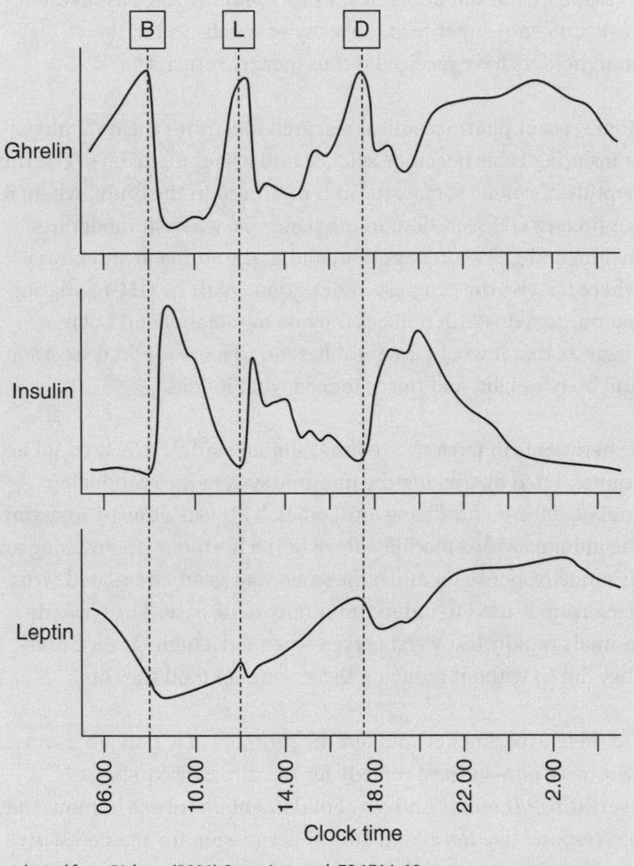

Adapted from *Diabetes* (2001) Cummings et al. **50** 1714–19

thermodynamics, either raise body temperature and heat loss (by sweating) or increase energy expenditure through extra physical activity. Similarly, somatostatin's multiple actions suggest that suppressing it with antibodies could cause stomach upsets and diarrhoea. It might also wreck the fine control of glucose levels by insulin and glucagon. Such side effects, especially if irreversible, could be devastating and overweight people might not be prepared to accept them.

Thus research in this area is difficult, potentially risky, and extremely hard to exploit. Nevertheless, because obesity and its related metabolic disorders account for so much morbidity and mortality, it has to be worth persisting. Studies on hormones often provide valuable insights into mechanisms. Whether they also become the basis for treatments and cures is quite a different question.

Chapter 7
The thyroid gland

Getting it in the neck

The Ukrainian nuclear power plant at Chernobyl exploded and caught fire on 26 April 1986 and released a radioactive cloud of fine particles and gas. Much of this material landed in adjacent parts of the western Soviet Union including Belarus and Russia, causing severe environmental contamination. Lighter material was blown further afield in two arcs, one northwestwards over Scandinavia and the other westwards over the Alps and parts of southern Europe.

About a fifth of the radioactivity came in the form of an isotope of iodine, I-131, and this proved useful for tracking the spread of the contamination. Iodine-131 produces a tenth of its radiation as gamma rays (the rest is beta radiation), which makes it easy to detect with a Geiger counter. Its half-life of eight days means that only a thousandth of it remains after 3 months, but this was long enough to allow an international army of environmental surveyors equipped with hand-held monitors to follow where it went.

One of the places they looked was in the necks of sheep, where the thyroid gland is located. The thyroid takes up iodine (whether radioactive or not) from the blood and stores it in a concentrated form attached to a large protein. Sheep grazing on contaminated land in the spring and summer of 1986 accumulated the I-131 and immediately became four-footed repositories of valuable environmental evidence.

Iodine is not found anywhere else in the body: the thyroid gland is unique in making use of it. This fact is exploited medically, not just for monitoring contamination but as a way of providing focused radiotherapy and tracing the delivery of drugs used to treat thyroid diseases. Unfortunately, the beta radiation from I-131 is itself carcinogenic, so other isotopes with slightly different radioactive characteristics (I-123 and I-125) are often preferable.

The thyroid is an unusual gland. Its cells are arranged in hollow balls, filled with the iodine-loaded protein called thyroglobulin. When stimulated by a hormone from the pituitary gland, thyroid cells build up the store of thyroglobulin but also turn some of it into a couple of hormones called thyroxine and triiodothyronine. We can call these T4 and T3, respectively, because the subscripts show the number of iodine atoms on each molecule. They have strange structures: besides containing iodine, each has two rings joined by an oxygen atom (Box 2). The rings come from the amino acid tyrosine, which is one of the main building blocks of the storage protein. T3 is more active than T4, and many tissues in the body, especially brain and fat, have an enzyme that strips T4 of one of its iodines.

Metabolic adjustment

In the second chapter, we considered the idea of a target organ—the place where a hormone has its action. The thyroid hormones illustrate just how misleading the word 'target' can be, for they work on practically every cell in the body. They speed up a cell's use of oxygen and so their effect is to accelerate metabolism in most tissues and organs. Whatever a particular cell happens to be doing, especially if it involves using energy to convert one biochemical substance into another, it will do more of it when T4 and T3 are around.

This general increase in metabolism extends to processes like immunity, growth, and reproduction, which is why the right level of thyroid activity is important for health. The thyroid hormones also make more energy available for muscular work and encourage the generation of heat, which can be crucial for effective thermoregulation. In animals, thyroid activity varies with the season and changes with day length, helping to explain many characteristic aspects of lifestyle such as seasonal breeding, migration, moulting, and hibernation.

The effects of thyroid hormones are widespread but not fast or dramatic. It takes a while for cells to respond, but the hormones have a correspondingly lengthy existence in the body. The half-life of T4 is about seven days, meaning that half of what you make today will still be in your blood this time next week. T3 has a somewhat shorter half-life but T4 can be converted into it so this makes little practical difference. The unusual structures of the hormones make them highly insoluble in blood, so in order to circulate they have to be attached very tightly to proteins. This also protects them and explains their impressive lifespan.

Thyroid disease

Thyroid diseases are relatively common medical conditions. They are also relatively easy to detect and often easily treated. The widespread effects of the hormones mean that sufferers may consult their doctors with a wide range of symptoms and some medical detective work may be needed to pinpoint the source of the problem.

The effects of an overactive thyroid, called hyperthyroidism, largely result from the increased metabolism. They include weight loss, increased appetite, sweating, warm skin, an intolerance of heat, and raised blood pressure, but may also include 'nervousness' and tremors.

On the other hand, severe thyroid inadequacy in adults produces a disease called myxedema. The metabolic rate is halved, hair thins and falls out, the skin develops a characteristic pallor, and mental activity slows down. More minor deficiencies can manifest themselves through reproductive, immunological, psychological, and other disturbances.

Somewhat similar symptoms to deficiency can result from thyroid hormone resistance. Because of their structure, thyroid hormones produce most of their effects by entering cells, binding to receptors, and changing the rate at which metabolism-related genes are expressed. There are several types of receptor but one of them, called TRβ, is particularly prone to absenteeism. This produces the symptoms of hypothyroidism even though the levels of hormone in the blood are normal. This deficiency is common amongst children with attention deficit hyperactivity disorder, perhaps reflecting a particular need for the TRβ receptor during neurological development.

Development and/or growth

The developmental effects of thyroxine have been known for over a century. In 1910, a New York scientist, J. F. Gudernatsch, went to work in Naples. He took with him extracts of thyroid, thymus, testicle, ovary, pituitary gland, adrenal gland, pancreas, liver, and rectum. Despite problems keeping the extracts cold in the spring heat of the Italian lab, he was able to test their effects on the development of fish and frog eggs. All of them disturbed egg development in some way and eggs died at varying rates over the next three weeks.

Gudernatsch needed clearer results, so the following year, now in Prague, he tried feeding tadpoles directly with small pieces of tissues taken from dogs and cats. The tadpoles ate these 'ravenously'. His control tadpoles were not fed but survived on 'tap water which in Prague is very rich on [*sic*] organisms'. Not surprisingly, the tissue-fed tadpoles grew twice as fast as the controls.

More interesting was the observation that those fed with thyroid gland died first, as dwarf but fully developed frogs. Gudernatsch noticed that all the animals in this group developed together. For example, they all grew fore limbs on day five and hind limbs on day nine. Development in the other groups was much less predictable, as well as taking a good deal longer. He also found that once tadpoles had been given thyroid gland, the metamorphosis could not be stopped by swapping to other materials.

Frogs are still widely used for studying early animal development. Nowadays, results are more likely to be in the form of data on gene expression and the timing of the appearance of particular proteins. Genes and proteins related to stages of limb and muscle formation, to the appearance and disappearance of gills and to the loss of the tail are particularly helpful in understanding what's going on and these developmental processes are now especially well understood.

One thing Gudernatsch's experiments showed clearly is that growth (making the body bigger) and differentiation (making different structures) are incompatible activities: parts of the body cannot do both at the same time. This is now an established principle of developmental biology. Growth depends on cells dividing and multiplying, whereas differentiation needs cells to change function, to move from one place to another, or to die in a controlled manner (called apoptosis). Crucial genes need to be switched on and off at crucial moments.

Thyroid hormones can affect all these processes but in different parts of the body at different times. As an embryonic animal develops, its thyroid gland needs to grow and make its hormones. In addition, potential target cells need to make hormone receptors. One of the main developmental genes affected by thyroid hormones is called 'sonic hedgehog' (named after a video game character, for reasons that need not detain us). It controls body segmentation, limb formation, and the development of organs including the brain and exists right across the animal kingdom, from fruit flies to fish and sea urchins to mammals. In amphibians, thyroxine-activated sonic hedgehog stimulates limb buds and shortens the tail, turning the tadpole larvae into adults.

Thyroxine has most of its growth effects early in life, during uterine development and infanthood. Low levels at crucial times can produce short stature as well as permanently impaired brain development and slow metabolism. In humans, low levels during foetal and neonatal life, either because the baby's thyroid is inadequate or because the mother's iodine intake is low, can be particularly unfortunate. The severe condition used to be called cretinism but is now more properly referred to as congenital hypothyroidism. Drastic effects on brain development can also happen if the foetal thyroid gland is malformed or when anti-thyroid antibodies produced by the mother's immune system cross the placenta. Some 20 million people worldwide are thought to be affected by thyroid-related developmental problems of one kind or another.

Thyroid deficiency diseases can be prevented. The idea of prophylaxis by supplementation started with the pleasingly named David Marine (1880–1976), an American pathologist who noticed that users of sea salt, which is naturally high in iodine, tended to be free of a neck swelling called goitre. Since then dietary supplementation through salt and other routes has greatly reduced the prevalence of deficiency diseases in most well-nourished societies. Nevertheless, some experts still consider low iodine intake to be a worldwide problem. A recent view is that sub-clinical iodine deficiency may be more common than previously realized, and that children born to mothers with an inadequate intake may go on to have lower IQs and do less well at school. Simple screening during pregnancy, followed by iodine supplementation, hormone treatment, or simple dietary advice (dairy and sea foods are good sources), should be able to eliminate this problem in societies with adequate healthcare.

Goitre

The secretion of hormones from the thyroid gland is stimulated by a large hormone from the anterior pituitary gland called TSH. The secretion of the TSH itself is stimulated by a tiny hormone from the hypothalamus called TRH (Chapter 2). Thus the production of the thyroid hormones is really the end of a chain of events controlled by the brain. Negative feedback—the control mechanism that results in self-regulation and homeostasis—is beautifully illustrated by this system: T3 and T4 reduce the secretion of TRH and TSH and so suppress their own stimulation. The activity of the brain depends on many things, both inside and outside the body, and has overall charge of the system, but the feedback means that there is constant self-regulation and adjustment.

Medical and physiological textbooks invariably illustrate how this axis works by showing (with alarming photos of people with distended necks and staring eyes) what happens when it goes wrong. The symptoms were described in Ireland in 1835 by Robert Graves (1796–1853) and independently in Germany in 1840 by Karl von Basedow (1799–1854). To this day it is called Graves' disease in the UK and America but von Basedow's disease in much of mainland Europe. By rights, both should have been beaten to eponymous posterity by Caleb Parry (1755–1822) who observed the condition in England 1786, or even by Sayyid Ismail al-Jurjani who seems to have noted it in 12th century Persia.

None of these individuals could have imagined the secretory function of the thyroid, let alone the hormones involved. Yet their careful observations established clear associations between the swelling of an organ in the neck, a strange bulging of the eyes and all the other characteristic symptoms that we now assemble under the heading of hyperthyroidism. Former US president George Bush Snr and his wife Barbara are both reputed, somewhat against the odds, to have been sufferers.

Graves' is an auto-immune disease in which the receptors for TSH on the cells of the thyroid get blocked by an antibody. Unusually for conditions of this kind, rather than disabling their target, Graves' antibodies permanently stimulate the receptors they bind to. The result is continuous secretion of thyroid hormones and growth of the gland itself. The hormones have their normal feedback effect on TSH and TRH, whose blood concentrations sink to zero, but that makes no difference to the activity of the gland. In a similar disease, Hashimoto's thyrotoxicosis, the antibodies may eventually destroy the gland completely. Both diseases are treated either by thyroid surgery, by thioureylene drugs that stop iodine being attached to the hormone, or by the radio-iodine treatments mentioned earlier.

A goitrous neck swelling with a different cause has been recognized for a good deal longer than Graves' disease. It was certainly known in ancient Greece and Rome, and there are many ancient, classical, and renaissance art works that can be interpreted as depicting sufferers of this condition.

It has been endemic in areas of the world with very low levels of iodine in the environment, including parts of Europe, West Africa, Australasia, South Asia, and the Great Lakes area of North America. In one particular region of the English Midlands it was known as Derbyshire Neck (reported as recently as the early 20th century: photographs on the internet are not difficult to find). Treatments based on seaweed, sea urchins, and other ocean products are described in early Chinese medical records and have been used in European medicine since at least the 12th century CE.

This type of goitre is due to lack of iodine and is associated with the symptoms of hypothyroidism (deficient thyroid activity). It therefore presents the apparent paradox of a large gland associated with a lack of the hormone it normally produces. The explanation lies in the loss of homeostasis: if iodine is missing, the thyroid can't make the hormone and there is no feedback on the secretion of TSH and TRH. The TSH continues to stimulate the thyroid and it grows. Regrettably, some otherwise useful textbooks and websites still portray this as the thyroid 'trying', ever harder and in vain, to gain more iodine. This is wrong: endocrine glands do not try to do anything, any more than the kidney tries to produce urine or the heart tries to pump blood. The cells of the thyroid gland simply do what their genes allow and respond to the conditions around them. Iodine deficiency goitre is what happens when negative feedback is missing.

Chapter 8
Light and dark

The rhythms of life

Writing in the journal *Science* in 1965, the German physician and physiologist Jurgen Aschoff said, 'Whatever physiological variables we measure, we usually find that there is a maximum value at one time of day and a minimum value at another.' Aschoff was pointing out the obvious but frequently ignored fact that the body is not a static object. Everything happening inside it fluctuates in speed or intensity, and does so in a more or less rhythmical way.

Many of the body's rhythms reflect the circadian (day/night) cycle. We see this in the way we sleep and our levels of alertness, in appetite, in body temperature, in our sensitivity to cold, light, noise, and other events around us, and even in our susceptibility to disease. We also experience more extended rhythms, over a year or over a lifetime, and can observe them especially clearly in the way some animals behave, reproduce, and survive.

Aschoff's insight came from studying his own body cycles but also those of volunteers who spent time cut off from the outside world in carefully regulated underground bunkers. He wanted to know whether the rhythms of physiology come from within the body or are driven by changes in the surroundings. He removed as many environmental signals as possible, leaving his subjects to control their conditions in whatever way felt natural to them, including whether to have the lights on or off and the timing of meals.

At around the same time (1960–70s) a French caver, Michelle Siffre, observed his own reactions and those of experimental subjects to living for up to four months in an underground cave. He was interested in the psychological effects as much the physiological ones, thinking that this might inform those facing the challenges of extended space flights that were being considered at the time. On returning to normal life, Siffre's subjects were perplexed to find that more days had passed than they had thought, and even felt that tricks had been played on them. As much as two months seemed to have been compressed into one.

The important conclusion that Aschoff and Siffre both came to was that the body has its own underlying rhythms but that these 'free run' in the absence of constant adjustment from outside. Left unchecked, the bodily cycles continue but they start to stretch, increasing in duration by a few minutes each day. In other words, the body's natural 'clock', whatever that might be, has a natural period slightly longer than that of the Earth's rotation.

The internal signals that misled Aschoff and Siffre's subjects about the passage of time were usually to do with sleep patterns and hunger. Aschoff's extensive observations and careful measurements demonstrated time (or phase) shifts in several underlying physiological cycles. These included the rise and fall in body temperature, rapid eye movement sleep, urine production, and the concentrations of hormones such as growth hormone and the adrenal-regulator ACTH. Thus his subjects were clearly experiencing more than just a psychological misperception. Aschoff and his colleagues were first to understand how environmental cycles keep the body's cycles in check, coining the word 'Zeitgeber' ('synchronizer' or 'time-giver') to identify reference events such as the transition between light and dark.

Chronobiology

The questions being asked by Aschoff, Siffre, and many investigators since turn out to have extremely complex answers. They make up the science of chronobiology, a rigorous academic discipline involving carefully controlled experiments and the objective interpretation of data, often with complex mathematical analysis. (Chronobiology is not to be confused with the pseudo-science of biorhythms that seeks to base all variations in the human condition on underlying cyclic phenomena, often in an anecdotal manner. There is a connection but they relate to one another in much the same way as astronomy relates to astrology.)

The rhythms we experience and that we can measure through hormone levels, metabolic changes, and body reactions are the product of several interacting influences. The 24-hour rhythms of light and dark are predictable but vary over the year and according to where we live. Environmental changes in the season, including food and water availability, can be crucial in the lives of many animals. The body also changes in age, maturity, and size, and has reproductive cycles of one kind or another.

It is very hard to sort out how these cycles interact, even using experimental animals in carefully controlled conditions: What controls what? How much is innate? How much is entrained? It is also challenging to work out how the body picks up and interprets the information provided by Zeitgebers. Current research shows that we have to take account of a network of systems involving the senses, parts of the brain, the activation and inactivation of genes, biochemical reactions, and cellular responses, all coordinated by complex hormonal cycles.

Melatonin

Some hormones reveal the day–night rhythm in a particularly dramatic way. Perhaps the most obvious is melatonin (Box 9).

Box 9. Melatonin

Approximate shape of blood melatonin concentration in sheep housed under conditions representing long day:short night (16 hours Light:8 hours dark) or short day:long night (8:16). This represents what happens in most mammals, including humans, over the year in non-equatorial latitudes.

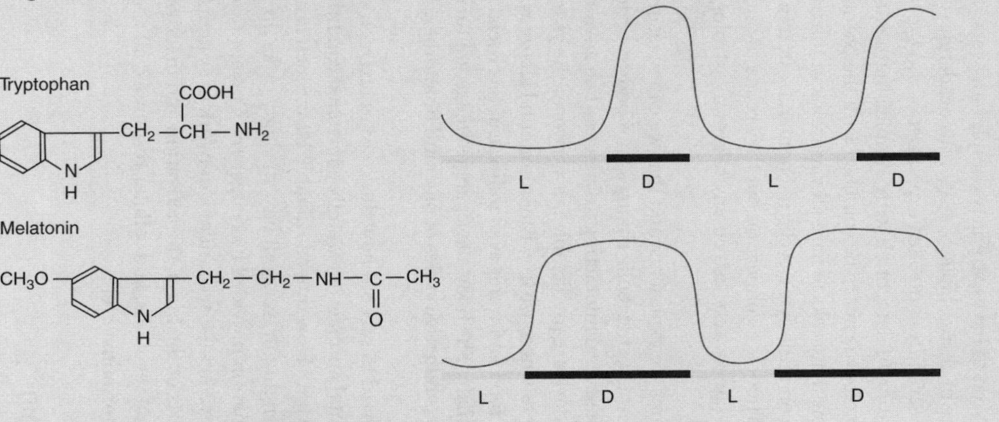

Melatonin is made by the pineal gland from the amino acid tryptophan.

The concentration of melatonin in the blood of sheep can range from nearly nothing during the day to 200–300 pmol/l during the night.

This hormone was discovered in the 1920s and named for its ability to cause skin darkening, first in frogs and then in other non-mammalian vertebrates. It comes from the pineal gland, a small structure nestling between the brain's cerebral hemispheres. The 17th century French philosopher Descartes thought the pineal gland was the seat of the soul. Whilst this is a difficult concept to test, we now know a great deal about how the pineal is controlled, how it connects to other parts of the brain, and the effects of the melatonin it produces.

The main influence on pineal secretion is daylight. In mammals, the amount of this is detected by the eye. We are familiar with the rods and cones that allow us to see, but the retina has a range of light-sensitive cells some of which help to track the overall amount of light over the circadian cycle. In birds, fish, amphibians, reptiles, and possibly in mammals as well, photo-sensitive cells exist within the brain itself, catching the small amount of light that penetrates the skin and skull. In fact, one view of the pineal gland is that it is a specialized collection of such light-sensitive brain cells, stimulated both directly and indirectly.

In mammals, information about the daily amount of light finds its way from the eye to the pineal gland by a tortuous route. Nerve connections take it first to a region of the hypothalamus called the supra-chiasmatic nucleus (SCN). This is where the body's master clock resides. Exactly how the SCN counts down time is still something of a mystery (we know it involves the rhythmic switching on and off of genes, but we barely understand where, how, or why), but light/dark information from the eye is the main Zeitgeber that keeps it in train with day and night. From the SCN, nerve connections go to the back of the brain, pass a short way down the spinal cord to the superior cervical ganglion (a kind of neural relay station), and then up to the pineal gland.

'Daytime', as interpreted by the day/night synchronized SCN clock, suppresses melatonin secretion from the pineal gland (Box 9).

Thus the amount of melatonin in the blood essentially tracks darkness: concentrations are high at night and low during the day. Changes in secretion happen over hours rather than minutes, so away from the equator the yearly extension and compression of the dark period gradually adjusts the SCN clock. This produces a corresponding lengthening and shortening of the melatonin profile.

Melatonin and breeding

Many animals depend on the melatonin rhythm to time their breeding. For example, sheep need the lengthening nights of the second half of the year to jolt their reproductive systems into action: ewes start to cycle and ovulate while rams grow their testes and make sperm. The activated gonads secrete steroid hormones and mating behaviour ensues, usually in early autumn. Gonad activity declines again as the days lengthen after the winter equinox, but during this time the females are pregnant—for about five months—and lambs are born in the spring.

Because mating happens in the autumn, the sheep is sometimes called a short-day breeder but this rather over-simplifies things. It would be fairer to call them 'non-long-day breeders' if that wasn't so cumbersome. Sheep at the equator, where day and night are of roughly equal length, can breed all the year round. Sheep kept artificially on continuous light tend not to breed at all. Sheep held for extended periods in the dark or given year-round melatonin treatment also fail to reproduce. Most types of sheep need 12–13 hours of light per day for about a month before they will be sensitive to longer nights and extended melatonin. Sheep can be made to breed outside the normal time of year simply by giving them melatonin. It is a small molecule that is cheap to make and easy to deliver in the form of implants or gastric boluses, so it has been used commercially to manage the breeding of sheep and some other domestic animals.

The influence of melatonin on the timing of reproduction has been known almost since the hormone's discovery, but exactly how it worked was a mystery for a long time, especially as it didn't have much effect on the gonads directly. In the 1980s and 1990s, melatonin's receptors were located by labelling the hormone with a radioactive tag and injecting it into sheep. The animals were killed and thin slices of body tissues were laid on photographic film so that the destination of the tagged hormone molecules would show up under the microscope as black spots. The researchers expected most of the spots to be distributed around the brain, especially in the hypothalamus. Yet although there were some in the hypothalamus, by far the largest accumulation was in the pars tuberalis, part of the anterior pituitary gland whose function was unknown at the time.

As with all unexpected results in science, colleagues in the research community were initially sceptical: they needed to be convinced that the pars tuberalis really was the main site of melatonin action and not just a randomly sticky tissue. The result was eventually shown to be reliable and this part of the pituitary is now well recognized as having the greatest abundance of melatonin-sensitive cells. The density of melatonin receptors in the pars tuberalis fluctuates on a seasonal basis and there are a number of other genes in this tissue that are expressed in a cyclic way. Further research on the pars tuberalis has shown that it influences many hormone systems, not just the reproductive axis. It has been described as a kind of gateway to the rest of the anterior pituitary gland—an intermediate between the brain (the hypothalamus, the SCN, and other regions) and the pituitary's main secretory cells (producing LH, FSH, TSH, prolactin, ACTH, and growth hormone).

But that is not the whole story. The thyroid gland must also be closely involved in the way melatonin works because sheep from which this gland has been surgically removed do not lose their breeding condition after the winter solstice as they normally would. The duration of the nightly secretion of melatonin decreases as expected but business in the ovaries and testes carries on regardless. The absence of thyroxine is essentially allowing these sheep to breed throughout the year. Experiments with tiny thyroxine implants in the brain point to parts of the hypothalamus once again, but they also suggest that melatonin changes the sensitivity of these sites to thyroxine rather than the other way round. Some of the key receptors, cell messengers, and genes are still being worked out, and the melatonin story remains a complex and confusing one.

The reproductive response to shortening day length, as illustrated by sheep, is a common feature of animals living in northern or southern latitudes but a few animals respond to melatonin in the opposite way. The golden hamster, for example, breeds in the summer months (March–September) and has a very short pregnancy (16 days). As with other animals, the activity of its pineal gland tracks changes in day length but it relies on contracting rather than expanding periods of melatonin secretion to bring it into reproductive activity. Thus the hamster's response to melatonin is essentially a mirror image of that of the sheep.

Melatonin and physiology

The darkness-driven melatonin cycles of animals do not only affect reproduction. Many changes that happen on an annual basis depend on variations in the secretion of prolactin from the pituitary gland. Good examples are feather renewal in birds, the winter thickening of the coat in many large mammals, and large scale changes in behaviour such as migration and hibernation. As with the reproductive hormones, the secretion of prolactin is influenced by melatonin through the hypothalamus and the pars tuberalis.

Melatonin has short-term effects too. Melatonin is taken by some travellers to prevent jetlag and it is available in some countries as a treatment for insomnia. It certainly causes drowsiness, plus other sleep-related responses such as lower body temperature, yawning, and reduced concentration, although there seems to be no agreement about its general efficacy or whether it really is a useful treatment for sleep disorders. The possible mechanisms for these effects are incompletely understood, but that may be largely because sleep itself is a physiologically mysterious process.

Cycles, cells, and cancer

Some of the most exciting discoveries about melatonin are to do with its effects on cancer cells. In the lab dish, melatonin can slow the division of cancer cells or allow healthy cells to survive at the expense of cancerous ones. These kinds of effects have been seen with pancreatic, prostate, breast, endometrial, ovarian, skin, and colorectal cancer cells, so they are an obvious target for research towards new pharmaceutical and other treatments.

Another seemingly promising target for medical research is melatonin's action on the brain and central nervous system. It can protect nerve cells from damage by the free radicals (highly reactive molecules with unpaired electrons) that abound in the environment and it might help to prevent neurodegenerative diseases such as Alzheimer's. At the psychological level, researchers have wondered whether melatonin could play a role in depression, and particularly in seasonal affective disorder, but research so far has been inconclusive.

In terms of metabolism, events such as blood glucose concentration, fat deposition, and other energy-related process are all melatonin-sensitive. In fact, researchers believe that all cells in the body are directly affected by the hormone to a certain extent. Metabolism-regulating hormones such as insulin, ghrelin, and leptin have a circadian aspect to their pattern of secretion. Obesity, diabetes, and high blood pressure have been investigated as potential targets for melatonin treatment. However, as with the anti-cancer and neurological actions, it is one thing to do tests on experimental animals or in a cell culture dish. It is quite another to turn the results into effective therapies or preventative lifestyle advice.

Cycles in the adrenal gland

The oscillations of the SCN clock, synchronized by day length and other environmental variables such as temperature and food availability, provide a reference for the rest of the body. All body cells have internal rhythms based on the switching on and off of genes, but these are really secondary clock systems because they depend on the SCN for coordination. Without this central entrainment, cellular biochemistry would run free and the body would become chaotic.

The SCN has some direct connections to parts of the hypothalamus and, as we have seen, uses melatonin to synchronize the brain and the pituitary gland. But how does the rest of the body feel its influence? The most important mechanism is the hypothalamic–pituitary–adrenal (HPA) axis.

The steroid hormone cortisol is secreted by the cortex of the adrenal gland under the stimulation of ACTH from the anterior pituitary gland. The corticotroph cells that make ACTH receive their marching orders from the hypothalamus (Chapter 2) and ultimately from other parts of the brain, and therefore come under the influence of the SCN and melatonin. Cortisol affects the metabolism and biochemistry of virtually every cell in the body, so here is a perfect system for linking the cycles in cells to those in the environment.

Measurements of ACTH and cortisol in the blood reveal regular, predictable patterns over the circadian cycle. The pituitary secretes

ACTH in pulses, following pulses in the CRH produced by the hypothalamus. The cortisol pattern is generally similar to that of ACTH but with the peaks and troughs running 10–20 minutes behind. The overall concentrations of these hormones are more or less steady during the day. Levels falls in the late afternoon, decline further through the evening and reach a minimum around midnight. They then increase, peaking in the early morning just before waking. If we look at the pulses of hormone in the blood (this involves repeated blood sampling and is easier to study in experimental animals than in man), changes in amplitude largely account for the changes in average concentration.

The cyclic effects of cortisol can be seen in the central nervous system, in the heart and circulation, in blood sugar and fat levels, in the amount of water in tissues, in the ability of the body to regulate its temperature, and in the reactions of the gut—in fact in most of those cycles we are aware of as part of normal life. It is hardly surprising that clock-defying disturbances such as jet travel, sleepless nights, or moving shift work from day to night cause us so much upset.

The HPA axis and cortisol are also central to the response to stress and disease. They provide links to shock hormones such as adrenaline, to some of the growth factors that control the development and lifecycles of cells, and to a wide range of mechanisms to do with defence such as inflammation and immunity. Studies in the 1990s showed that ovarian cancer patients had a three to four times better chance of survival if their chemotherapy was delivered at 6.00 am rather than 6.00 pm. In other patients, bowel cancer drugs had fewer side effects if they were given at fixed times of the day rather than continuously infused. Statistics also show that night-shift workers have greater than average rates of cancer, neurodegenerative diseases, asthma, cardiovascular disease, and gastrointestinal disturbances. Perhaps their upset cortisol and melatonin patterns have a part to play in this.

Evolution

The melatonin/thyroxine and ACTH/cortisol mechanisms are examples of systems that coordinate events around the body. Using information from the SCN, they synchronize cellular biochemical clocks and account for the ubiquitous cyclicity of body systems to which Aschoff alerted us. Most other hormones, whether secreted by the brain, the pituitary, the adrenal, the gonads, or even the gut, show cyclic patterns of secretion and it would be fair to say that hormones are central to all aspects of chronobiology. An interesting question is why organisms should have evolved to be cyclic in this way. What advantage is provided by the ability to respond to changes in day length, temperature, food supply, and other features of the environment?

The value of synchronizing reproduction with seasonal factors is the easiest to understand. The birth of lambs in early spring in temperate regions results in them being first fed by their mothers but then weaned at precisely the time when grass reaches its highest nutritional value. Furthermore, growth and development during the summer mean that the yearling lamb will be robust enough to survive the subsequent winter, and may even be able to breed in its first season.

More subtle use of environmental signals allows animals to survive under challenging conditions. The winter torpor of many temperate species and the classic hibernation of arctic mammals such as polar bears is preceded by increases in appetite, food intake, and fat accumulation, all stimulated by decreasing day length. The impressive annual migrations of many bird species are tightly coordinated with the need to moult and renew plumage, as well as with their highly seasonal reproductive behaviour. Several species of equatorial bat use seasonal patterns of rainfall to time with astonishing precision their periods of hibernation, swarming, social interaction, and mating. Seals may migrate impressive distances to one particular breeding ground at a precise time of year, so that they can give birth and immediately mate again before heading back out to the oceans.

In these and many other examples, it is possible to interpret environmental sensitivity and the associated metabolic, physiological, and behavioural responses as *anticipatory* survival devices. Those animals that are best able to predict events in their environment will be best able to withstand adverse conditions, exploit abundant resources as they appear, avoid predators, and reproduce successfully. What we do *not* need to do is impute some conscious aspect to the anticipation—the animal is not required to think ahead. It is quite sufficient to explain the responses of environmentally sensitive animals, however complex and unlikely they may appear, as having evolved under the environmental pressure to survive. Those animals that changed in a certain way at a certain time of the year survived and reproduced; those that did not died out. Hormones, regulated by the brain and ultimately by external events, provide the internal coordination that enables this to happen.

Chapter 9
Changing perspectives

Taking a long view

Hormone science is well over a century old, as we saw in Chapter 1, but it continues to develop in rapid and exciting ways. Since the description of secretin in 1902, countless new hormones have been discovered and it would be a vain challenge to try to list them all. It might be reasonable to ask, 'Have we found them all yet?' Although the answer would have to be a firm 'No', we would need to be a little careful about what we meant. It is time to think again about what hormones really are and to re-evaluate the complexity of the biological systems they represent.

The traditional view of hormones, going back to Bayliss, Starling, and Bernard and encapsulated in Doisy's paradigm (Box 1), has been that they are individually identifiable chemical signals, secreted in one place in the body and acting in another, which facilitate homeostasis and make reproduction, development, metabolism, and survival possible. This view is informative and reasonable if what we want to do is describe, catalogue, and understand the body's internal control systems. Most physiological and medical textbooks take this approach, and I have largely adopted it in these pages too. But the information provided by contemporary molecular, genetic, and physiological research shows it to be an oversimplification.

With modern insight it is more realistic to see the growing list of hormones as a collection of biologically active chemicals, each of which just happens to cause one cell to change its behaviour in response to another. Any internally secreted chemical that does this can justifiably be called a hormone, even if it is not among the proteins, steroids, and other molecules traditionally described in endocrinology textbooks. We should also resist the temptation of seeing hormones and their receptors as making up the parts of carefully designed control systems, or at least of systems that follow a set of fixed anatomical and physiological principles. The difference in perspective is a subtle but important one. It is a bit like describing the internet as a structured network of information resources: in reality it is a random and uncountable collection of devices linked by ever-changing connections and operating by constantly evolving rules of engagement.

Complex hormones

The first hints of complexity came in the 1960s and 1970s. As more and more hormones were discovered, in different places in the body, and under different physiological conditions, they started to be recognized as belonging to families. Technical developments enabled smaller and smaller amounts to be measured and it became increasingly possible to identify tiny but significant differences in molecular structure. Three examples will illustrate the point, although many others can be gleaned from earlier material in this book.

As the steroid hormones were discovered and understood, those coming from the adrenal gland were found to be closely related to those from the ovary and testis. The relationship is structural—they all have a basic similarity to cholesterol—but also biochemical—they are really just different products of one large, complex metabolic pathway (Box 4). We might think of adrenal cells making cortisol, testis cells making testosterone, and ovarian cells making oestrogen, but in reality all three organs can make most or all of those hormones and a great many others as well. The best way to understand steroid hormones is not to consider what individual organs do but to look at which cells follow which parts of the metabolic pathway most actively.

As a second example, consider the strange origin of ACTH, the anterior pituitary hormone that stimulates adrenal cortisol secretion. It circulates in the blood as a protein of 39 amino acids but it starts life as a protein of 265 amino acids called POMC (which stands for pro-opiomelanocortin, should you really want to know). POMC is made in at least two parts of the pituitary, called the pars distalis and the pars intermedia, and also in the brain. Its destiny depends on the enzymes it meets: in the pars distalis, enzymes chop it into ACTH, beta lipotrophin (a fat regulator), beta endorphin (one of the natural opiate painkillers), and some other fragments. In the pars intermedia, slightly different enzymes chop it into two types of melanocyte stimulating hormone (which change skin and fur colour), some endorphin-like peptides and other fragments. So although there is essentially one gene and one starting material, the products depend on location. POMC-like proteins have been found in all animals with spinal cords. The list of functions carried out by its products is vast and different animals use them in different ways.

A third example is the family of hormones called the glycoproteins. These hormones are also secreted from the anterior pituitary gland but they have a completely different structure to ACTH. Each comprises a pair of large interlocking proteins, enveloped in a big blob of starchy gel. They include the gonad-stimulating hormones LH and FSH and the thyroid stimulator TSH. They form a family because one of the interlocking proteins, designated alpha, is the same in all of them. To this family we can add the chorionic gonadotrophin or hCG (h for human), which we met in Chapter 3. Although this comes from embryonic tissues early in pregnancy so doesn't normally exist in the mature body, it has a similar structure and the same alpha protein as the pituitary hormones. But the family doesn't really stop there: each glycoprotein comes in several different variants (called isoforms). These have minute differences in protein structure—just one or two amino acids amongst thousands—but vary considerably in their effectiveness as hormones.

Complex systems

Before trying to interpret what all this means, we should extend our appreciation of endocrine complexity a few steps further.

As explained in Chapter 2, the receptors and internal cell messenger systems through which hormones have their effects on cells also fall neatly into families. Within these groupings, the basic structures and mechanisms are often similar, but quite subtle differences can account for enormous variability. This is well illustrated by the most common hormone receptor family, the G-protein-coupled receptor or GPCR family. This is often called a superfamily because its many hundreds of members can be divided into at least five distinct classes. Amongst these are variants that respond to hypothalamic neurotransmitters, to most of the hormones from the pituitary gland, to adrenaline produced during shock, and to parathyroid hormone released when calcium levels fall (to mention but a few examples). GPCRs also play a central role in the immune system, in the brain's response to drugs and anaesthetics and in the senses of vision, taste, and smell. Examples of GPCRs can be found in worms, fungi, and flowering plants, while others allow chemical communication between single celled organisms. In understanding how hormones work, we are clearly dealing with just a small representation of a widespread biological phenomenon.

To appreciate hormone and receptor families properly we should really examine the genes that encode them. Modern methods let us do this very easily, revealing the DNA sequences that either underlie the structures of the hormone and receptor proteins themselves or encode the structures of the enzymes that make them. This information shows the true nature and extent of the families and can often throw up unexpected relationships between familiar but superficially distinct groups of hormones. The human GPCR family has been estimated to be the product of some 800 genes, perhaps accounting for as much as 4 per cent of the whole genome.

If we zoom back out to the level of the whole body and the systems that coordinate its physiology, hormone complexity takes on a different form. In earlier chapters we described how many hormones, notably those of the reproductive system and the brain–pituitary–adrenal system, appear in the blood in short busts. Their pulsatility is described as ultradian, meaning that the oscillations are measureable in short time frames (minutes or hours) rather than the longer periods associated with daily (circadian) or yearly (circannual) rhythms.

Pulsatility is far from just a curious observation. Mathematical modelling shows it to be more energy efficient than continuous secretion and also more resistant to disturbance. As with radio communications, the information value of the signal can be increased by varying pulse amplitude ('AM'), frequency ('FM') or shape, whilst the gaps between pulses allow target cells to recover, making them more responsive.

The cellular and biochemical origin of pulsatile secretion is imperfectly understood. The pulsatility of the pituitary hormones depends partly on a 'pulse generator' in the hypothalamus, partly on how the hormones are synthesized, partly on negative feedback from hormones further down the system (from target glands: ovary, testis, adrenal, etc.), and partly on how biochemical processes *inside* the pituitary and target gland cells behave. Unravelling this requires the combined efforts of molecular geneticists, biochemists, physiologists, mathematicians, and statisticians, so it is not surprising that progress is slow.

Despite the mystery surrounding its generation, the physiological importance of pulsatility is clear. We have already seen how changes in the amplitude of ACTH pulses contribute to different levels of adrenal gland activity between morning and evening and therefore to the overall rhythms of the body and its metabolism. In the reproductive system, the frequency of hypothalamic and pituitary hormone pulses changes over the menstrual cycle (one pulse every 90 minutes during the follicular phase, compared to one every six hours at most other times). At puberty, an increase in pulse frequency drives the development of eggs and sperm and the concurrent increase in gonadal hormone production. Given that species survival depends on reproduction, some physiologists go so far as to argue that hormone pulsatility is fundamental to the continuation of life.

Hormone evolution

A good way to make sense of all this and to understand what hormones really are, is to take an evolutionary perspective. In 1982, Hugh Niall set out four rules to guide us (Box 10). They are a reminder that animals are biological entities that have arisen through evolution, not by design. Evolutionary change happens through random variation coupled with the selection of individuals according to their ability to survive and reproduce. The basis of the variation is unplanned mutations, or mistakes, in genes. These happen in all genes on a slow but statistically regular basis. Mutations result in changes in proteins. Most of these are irrelevant but just a few are either lethal or advantageous to the individuals that possess them.

Niall's Rule 1 explains that chromosome replication during cell division sometimes goes wrong, leading to cells with more than one copy of certain genes. Suppose that an individual animal has a duplicated hormone gene and that one copy develops a mutation while the other does not. Suppose too that the mutated copy makes a variant of the hormone that happens to help the individual survive

Box 10. Niall's Rules of hormone evolution and Luck's lemmas

Rule 1 'Gene Duplication is the name of the game'

Rule 2 'Everything is made everywhere'

Rule 3 'Never make a new hormone if you can use an old one'

Rule 4 'Conservation of structure = function'

Niall proposed these friendly rules as a guide for understanding the evolution of peptide hormones. They apply equally well to the evolution of receptors, binding proteins, growth factors, and other proteins in the hormone story, as well as to the enzyme systems that make non-protein hormones like steroids, thyroxine, and adrenaline. In fact, they apply to the evolution of everything in biology.

From 'The Evolution of Peptide Hormones', *Annual Review of Physiology* (1982) Niall HD, **44** 615–24

Luck's lemmas:

Rule 3a 'Don't be fooled by the labels you put on things'

Rule 4a 'Selection acts on phenotype, not genotype'

It is presumptuous to add to Niall's inspired insight, but these reminders may help with interpretation. They, too, apply to everything in biology, not just hormones. An organism's phenotype is all of its observable, physical characteristics; its genotype is its personal collection of genes.

and reproduce. That individual's offspring, which have the benefit of two useful hormones instead of one, will spread their genes through the population. If this scenario happens many times, it is not hard to see why families of hormones and receptors should emerge. Minor variants become valuable if they give an organism an advantage over others or allow it to survive and reproduce under new conditions.

Rule 2 reminds us that, with the exception of eggs and sperm, all the body's cells have identical genes. So any cell can make any protein. The difference between, say, a pituitary cell and an adrenal cell lies in which genes are most highly expressed. Niall's rule exaggerates slightly (because genes can in fact be switched off completely) but the principle is correct. It means that we shouldn't be surprised when hormones are found in more than one place (somatostatin in the pancreas and hypothalamus, neuropeptides in the gut and brain, glycoproteins in the embryo and pituitary). It also encourages researchers to look for hormones of all kinds, not just the ones they expect, in whatever tissue they happen to be studying. 'Never say never' is a wise mantra in hormone research.

Rule 3 sounds like one of 'Nature's' commandments, but it expresses a profound truth: evolution is opportunistic and frugal. Organisms survive, reproduce, and spread their genes by exploiting whatever attributes they have available. Assigning a name to a hormone is a human conceit. What we first discover that it is doing reflects our interests, the progress of science, and the organisms we happen to be studying: it does not define some invariant, intrinsic, or intended property of the hormone (my Rule 3a). Physiological control systems do not come ready made, nor are hormones invented to make them function; they emerge by adaptation on the basis of what is available and what works. Niall invokes the wisdom of Sir Peter Medawar: '[Hormone] evolution is not an evolution of hormones but an evolution of the uses to which they are put.'

Rule 4 is another truism. If a hormone works, promoting an animal's survival and reproduction, the gene for it will be passed on to the next generation. All genes experience random mutation (see Rule 1), so genes that do not code for useful variants will gradually accumulate errors and fade into the background DNA. However, evolutionary selection does not work on genes: it works on the characteristics (called the phenotype) resulting from the proteins that genes code for (my Rule 4a). This means that anything we discover (proteins, enzymes, hormones, and everything else) must be doing something useful. This applies within the organism, between organisms, and between species. For example, the posterior pituitary hormone oxytocin is also made by the ovary in some animals and there are even oxytocin-like proteins in earthworms and hydra. Oxytocin is therefore described as being highly conserved across evolutionary time. It must do something very useful, but that use has been adapted for a myriad different circumstances.

With this evolutionary perspective we can now understand how the massive complexity of hormones and endocrine systems has emerged. It is also evident that hormones and their receptors must have evolved together, otherwise they would not work. Yet this is not as remarkable as it sounds if we recall that evolutionary change happens not by the direct selection of individual genes or proteins but according to the survival and reproduction of the cells and organisms that possess them.

Are hormones still evolving? Almost certainly. But where did the first hormone come from? We already have the answer: cells respond to chemicals and the chemicals we call hormones happen to be those released by other cells. Hormones began as soon as cells started to cooperate through chemical communication. Many chemicals besides the traditional hormones have this role: neurotransmitters (which signal between nerves and between nerves and muscles), cytokines (which signal between immune cells), pheromones (by which individual animals communicate), metabolites, and excretory products.

Trying to work out which of these became the first hormone is an academic point and does not really get us anywhere. Any cell that produces a chemical signal, wherever it is located, is an endocrine cell, and all cells communicate using chemical signals. All cells are endocrine; it is just that some are more endocrine than others.

Medical exploitation of hormones

Endocrinology is a major field of medical research, as indicated in various places in this book. There are abundant examples of how contemporary understanding has been exploited to detect, prevent, treat, and cure disease. Treatment may involve direct intervention such as the surgical removal of a malfunctioning gland or a hormone-secreting tumour. It may require injections of natural and synthetic hormone preparations, or the use of pharmaceuticals that replicate or block hormone actions. The medical profession takes the existence of these therapies for granted and knows that each one is supported by evidence of efficacy.

There are, however, many medical conditions that have a clear hormonal involvement but that seem stubbornly resistant to simple practical intervention. Good examples are the control of appetite, obesity, Type 2 diabetes, and related diseases in humans, and the similar metabolic syndromes frequently found in domestic animals such as dogs and horses. It currently seems possible to understand individual elements of these diseases but the complexity of the whole condition can be daunting and bewildering. Nevertheless, with continued research, especially if it leads to individually tailored genetic and physiological information, we can be optimistic that the efficacy of therapeutic interventions will continue to improve.

The ethics of hormone use and abuse

As with most medical and physiological research, studies on hormones and their application attract a number of ethical questions. Some of these are rather controversial and hard to resolve. They concern the way in which information is gained and the use to which that information is put.

Animals have been used for hormone experiments right from the start of the subject's history. This exploitation continues although current research is much more likely to entail genetic manipulation, biochemical investigation, and tissue analysis rather than the clumsy gland removal and extract injection approaches of the distant past. Animals remain important subjects of study because they are really the only effective context for investigating how complete physiological systems work. Individual cells in a culture dish, gene sequence analysers, and hormone assays can provide valuable information on how cells respond, how secretions are controlled, and the effects hormones have. They are also helpful in studying receptor systems, biochemical processes, and metabolic pathways. Yet they offer limited scope for studying cell–cell communication on a whole organ or body scale.

Hormone researchers, like other biological and medical scientists, use animals when they have to—not because they want to. They have no interest in causing animals to suffer and they do not enjoy inflicting pain or distress any more than anyone else. The ethical question is whether society feels this type of research is justified in terms of the knowledge gained and its potential application. The ethical debate rightly continues in the public square. Researchers, scientists, and doctors may be better informed about the technicalities, but their views carry no greater moral weight than those of anyone else.

Researchers experience the impact of this debate in the amount of funding made available for the studies they wish to do. They also have to comply with strict procedural regulations and the completely reasonable, if sometimes tiresome, requirement that they justify the use of every individual animal. Over time, society decides what it will and will not accept, and sometimes science itself can offer new ways around troublesome questions. Fortunately, the days of unregulated mass animal usage, to identify new hormones and measure their concentrations in research and medical samples, are long since gone. At the present time, genetic manipulation leads the way but experiments are still tolerated on some animals (invertebrates, rodents) more than others (primates, humans).

Another major ethical debate of our time is about the use of hormones to improve human and animal performance. One end of the debate, about which there is little controversy, concerns the use of hormones in sport. High profile cases such as sprinter Ben Johnson's use of anabolic (muscle-mass-increasing) androgenic steroids, cyclist Lance Armstrong's use of various doping materials (growth hormone, steroids both natural and man-made, and a synthetic version of the red blood cell stimulator erythropoietin), and even the injection of anabolics into racehorses and greyhounds are universally regarded as cheating.

Despite the millions of dollars spent around the world on monitoring programmes, it has become increasingly difficult to detect such practices. This is partly because modern drug preparations closely mimic naturally occurring hormones and partly because users have become adept at hiding them, for example by aligning treatments to underlying circadian variations in the body's own secretions or making them components of acceptable medicines. There is essentially a technical arms race between, on the one side, sport's guardians of 'fairness', their laboratories, and their ever-increasing list of banned substances, and on the other, clever pharmaceutical chemists, well-informed coaches, and their highly rewarded protégés. In a purely factual sense, both protagonists have been impressively clever in making use of everything we know about hormones and how they work.

At the other extreme, as we have seen, society finds it perfectly acceptable for hormonal treatments to be used to prevent and treat disease. Indeed, they improve the quality of life for very many people. We would be horrified if, because of a moral objection to 'doping', diabetics were denied insulin, children of restricted stature were denied growth hormone, or mothers experiencing slow delivery were denied contraction-inducing oxytocin.

The fact that the hormones we now use are produced by chemical and molecular means, rather than by extraction from animal or human cadavers, scarcely alters the point. Brown-Sequard and Voronoff first used glandular extracts in attempts to rejuvenate themselves and stave off the effects of ageing. To us, with the hindsight of more than a century of further research, these pioneering endeavours and their modern variants seem physiologically clumsy and medically misguided. Yet it would be interesting to observe society's reaction if methods with equivalent novelty and apparent promise were proposed today in other areas of medical or pharmaceutical science.

Perhaps the closest we can get to sensing the potential depth of controversy in our own times is to recall reactions to developments in reproductive medicine the second half of the 20th century. We have already noted the religious criticism of IVF and related techniques. A related example is the reaction to the emergence of the contraceptive pill in the 1960s and 1970s. Hormonal contraception arose as the inevitable consequence of the emerging understanding of how hormones control the female reproductive cycle, coupled with the ability of chemists to produce the required chemicals on a large scale. Some groups saw all this as a grave threat to moral and social order, while others saw it as offering women liberation from reproductive tyranny.

Overall, there seems to be something about competitive performance, as opposed to medical treatment, which changes the ethical perspective. From a purely logical point of view, it is quite hard to find an absolute moral difference between the use of synthetic growth hormone to increase a child's height and the use of the same material to increase a weight lifter's muscle mass. Both supplements work by exploiting the body's genetic and metabolic potential and they both depend absolutely on adequate supporting nutrition, yet one is seen as correcting a deficiency and the other as giving an unfair competitive advantage. What if the treated child grew up to be an exceptional athlete?

Interestingly, in the case of drug abuse by athletes, opprobrium tends to focus on the cheating rather than on the potential side effects of the materials used. The latter aspect would seem to offer greater moral traction, for the aim is presumably to protect individuals from dangerous or socially misguided behaviours, to avoid setting inappropriate examples to the young, or to reduce the risk of unintended consequences. But even the safety argument loses some of its persuasiveness if we return to the example of the mass use of contraceptive hormones: all those ingested hormones have eventually to be excreted into the environment and are well-known to pose toxic dangers for humans and animals. This information is fully in the public domain but is rarely if ever used as an argument against contraceptive use.

These are value issues for society to resolve, not scientists, but it surely helps if arguments are based on sound understanding and a balanced presentation of what we know.

Where next?

If William Rice was born today, his condition would be diagnosed and evaluated within an endocrine paradigm, and his parents would accept unquestioningly his doctors' attempts to correct it. He would undergo repeated assessments as he developed, and steps would be taken to block, augment, or stimulate whichever aspect of his endocrinology was found to be disturbed. His doctors would have to hand the wisdom of a century of research on what regulates human growth, a rich palette of pharmaceuticals of proven efficacy, and the accumulated experiential knowledge of surgeons, physicians, geneticists, and psychologists who had treated people with similar conditions.

The most impressive aspect of young William's diagnosis and treatment would be its individuality. In endocrinology, as in many other aspects of medical science, so much is now known that doctors and scientists can step with confidence beyond their general understanding of how the body works to treat each patient personally for his or her exact condition. Both diagnosis and treatment can be based on assessments that penetrate to the cellular, molecular, and gene levels. William's doctors would, furthermore, understand that hormones alone may not explain his symptoms: they may need to evaluate hormonal interactions with the immune, nervous, digestive, circulatory, and renal systems in order to develop a complete picture of his condition.

Endocrinology as an academic discipline is generally well funded and productive, at least in those parts of the world where the biomedical sciences are viewed as suitable targets for public and charitable support. Numerous high quality research journals are devoted to hormonal topics, usually with some major system specialization (nutrition, metabolism, reproduction, development, bone, etc.), and many receive so many manuscripts for publication that they have the luxury of being able to reject all but the very best submissions. A glance through any of these publications reveals not only the astounding technical complexity of the research work but also the diversity of specialist interests required to make progress, including bioinformatics, mathematical modelling, and epidemiology as much as cell biology, molecular biochemistry, and DNA sequencing.

Endocrine research also proceeds apace in the veterinary and animal sciences, especially regarding companion animals and horses, productive farm animals, and endangered or exotic species. The diagnostic and pharmaceutical industries employ endocrinologists with a diverse range of expertise. Complementing and underpinning all of this is research with no immediate application: that which seeks to understand the fundamental principles of how hormones work and their place in physiology. A somewhat smaller but no less diligent biological specialism continues to seek answers at the level of the environment, the biosphere, and evolution. There are worthwhile scientific careers to be had in all these areas.

It would be pointless to try to predict where endocrine research will lead next, but there are a number of interesting targets currently in view. We have already mentioned the challenge of worldwide socio-medical problems such as metabolic syndrome, diabetes, obesity, and appetite control. Specialisms as diverse as oncology, fertility management, and blood pressure control have, as we have seen, hormonal components that can produce medically and socially valuable applications. Neuroendocrinology is a comparatively young discipline with the potential to influence our lives in unexpected but dramatic ways. The emerging field of epigenetics is bringing new perspectives to the endocrine regulation of development and will undoubtedly present significant possibilities.

We will also find ourselves challenged by unforeseen, controversial applications of hormone research, not only in sport and human performance but perhaps in the availability of new recreational drugs, in the manipulation of lifespan, in the selection of individuals with desirable characteristics, or in the optimization of education. We have become accustomed in the recent past to dealing with unprecedented possibilities emerging from genetics. We would be wise to prepare ourselves for some equally surprising developments in endocrinology.

Further reading

I hope this little book has stimulated your interest in hormones. If it has, you might want to find out more. Where you look for further information will depend on your interests, your background understanding, and the level of detail you want.

General information

Most general textbooks of biology, written for school and early undergraduate study, have introductory sections on hormones. Here are three examples, from the many available. (With all the books listed, find the latest edition if you can.)

Biology (2014) Solomon EP, Martin C, Martin DW & Berg LR (10th ed.) Brooks/Cole

Biology, the Dynamic Science (2013) Russell PJ, Hertz PE & McMillan B (3rd ed.) Cengage Learning

Campbell Biology (2013) Reece JB, Urry LA, Cain ML, Wasserman SA, Minorsky PV & Jackson RB (10th ed.) Benjamin Cummings

More extensive information on hormones and their place in coordinating how the body works can be found in textbooks of general physiology. The following have particularly clear accounts and straightforward diagrams, as do several others:

Human Physiology: An Integrated Approach (2014) Silverthorn DE (6th ed.) Pearson Education

An Introduction to Human Physiology (2013) Sherwood L (8th ed.) Brooks/Cole

Animal Physiology (2012) Hill RW, Wyse GA & Anderson M (3rd ed.) Sinauer Associates

Animal Physiology: From Genes to Organisms (2013) Sherwood L, Klandorf H & Yancey PH (2nd ed.) Brooks/Cole

Not all student-level textbooks are reliable in every particular. Many are good on some topics and less good on others, and some authors have a habit of reaching for unconsidered generalities or colloquial descriptions when dealing with matters outside their own areas of specialism. The best way round this problem is to consult more than one book.

Detailed textbooks

Medical and veterinary textbooks contain more detailed coverage of endocrinology. As you would expect, they usually direct their presentations towards the understanding of diseases but they usually provide comprehensive accounts. Particularly good are:

Ganong's Review of Medical Physiology (2012) Barrett KE, Brooks H, Boitano S & Barman S (24th ed.) McGraw Hill

Integrative Endocrinology (2013) Laycock J & Meeran K Wiley-Blackwell

Metabolic and Endocrine Physiology (2012) Engelking LR (3rd ed.) Teton NewMedia

There are several specialist endocrinology text books. Some of these present the subject from a physiological or system perspective whilst others focus on, say, the biochemistry or pharmacology of hormone molecules and their receptors:

The Endocrine System at a Glance (2011) Greenstein B & Wood D (3rd ed.) Wiley-Blackwell

Vertebrate Endocrinology (2013) Norris DO & Carr JA (5th ed.) Academic Press

Molecular Endocrinology (2004) Bolander FA (3rd ed.) Elsevier Academic Press

Endocrinology (2007) Hadley ME & Levine JE (6th ed.) Pearson Prentice Hall

Historical

The classic reference book on the history of the subject is:

The History of Clinical Endocrinology: A Comprehensive Account of Endocrinology from Earliest Times to the Present Day (1993) Medvei VC CRC Press

but regrettably this is now out of print and you may have to search a few libraries to locate a copy. It is worth the effort, however, because its stories of hormone discovery and profiles of the discoverers are highly readable as well as authoritative and detailed.

Academic research

In a fast moving scientific subject like endocrinology, textbooks are always out of date. If you want contemporary information you will have to peruse the academic research literature. The most respected journals in the field are *Endocrinology*, *Journal of Endocrinology*, and *Journal of Clinical Endocrinology & Metabolism* but there are several others.

Be aware, however, that hormone research is reported in the journals of many other academic specialisms (nutrition, reproduction, development, neuroscience, etc.). It is best to locate what you want by searching for a topic, rather than by subscribing and reading journal issues from cover to cover (a recipe for madness as well as penury). The best way to locate relevant articles is by entering keywords into a search engine such as *Google Scholar* or, if you have access to them, *Web of Knowledge* or *PubMed*.

The extreme efficiency of these search engines means that you are likely to get back more than you bargained for. You will need several attempts at refining and combining your keywords, but if this still generates overwhelming output, try adding 'review' as one of your keywords. This helps to throw up the papers which summarize the topic, either at a detailed level or in a more digestible form. You must still be prepared to wade through a lot of biochemistry, cell biology, genetics, pharmacology, and data, for this is the nature of the subject at a research level.

Generally speaking, the research journals make their content freely available in downloadable pdf format after a period of time (usually a year) has elapsed from the date of publication. If you want articles hot of the virtual press you may have to pay.

Review journals

Lying somewhere in between the accessible-but-out-of-date textbooks and the unreadable-but-up-to-date research literature are the review journals. These present overviews of specialist topics written by experts in the field. They are generally understandable by anyone with a reasonable appreciation of the field and an enquiring mind. Coverage of endocrine topics is sporadic but you will often find what you need, with most articles free online. Some are run by learned societies while others are prepared by special interest groups, charities, companies, or other organizations. In the latter cases, take cognisance of origin before assuming balance and objectivity.

The *Annual Reviews* series is a particularly rich (and trustworthy) resource, offering well-written and digestible articles. Its website <http://www.annualreviews.org/> has an excellent search facility. Useful titles for endocrinology include *Annual Review of Physiology*, *Annual Review of Medicine*, *Annual Review of Neuroscience*, *Annual Review of Nutrition*, and *Annual Review of Biochemistry*, but a keyword search will ensure that you don't miss anything.

General guidance

As with all physiological subjects, endocrinology does not really exist except insofar as it is an integrated feature of how the body works. Thus your search for further information may be most fruitful if you can identify what you are really interested in. You may wish to pursue major topics (obesity, water balance, reproduction, growth, cell biology, cancer, etc.), or more focused areas (appetite control, hypertension, pregnancy, pituitary disorders, intracellular signalling, thyroid tumours, etc.). Alternatively, your interests may lead you towards generic studies (receptor evolution, steroid hormones, gene regulation, environmental pollutants, etc.).

If you are not quite sure where to begin, Wikipedia can provide helpful overviews and point you to valuable resources. However, please remember that its articles have not been subjected to peer review (unlike those in academic journals and textbooks) and are written by enthusiasts who are not necessarily experts.

Health warning

Finally, and importantly, please note that the present book does not offer medically dependable information and its author is not a health professional. If you are concerned about a particular medical or health condition, whether in a human being or an animal, please consult an appropriately qualified practitioner.

Index

Hormone names are listed with initial capital letters and their commonly used abbreviations in parentheses.

C

D

E

I

J

K

L

M

N

O

P

R

S